Lecture Notes in Mathematics

Edited by A. Dold and B. Eckmann

639

Norbert Adasch
Bruno Ernst
Dieter Keim

Topological Vector Spaces

The Theory Without Convexity Conditions

Springer-Verlag
Berlin Heidelberg New York 1978

Authors
Norbert Adasch
Fachbereich der Mathematik der
Universität
Robert Mayer-Str. 10
D-6000 Frankfurt am Main

Bruno Ernst
Fachbereich Mathematik der
Gesamthochschule
Warburger Str. 100
D-4790 Paderborn

Dieter Keim
Fachbereich Mathematik der
Universität
Robert Mayer-Str. 10
D-6000 Frankfurt am Main

AMS Subject Classifications (1970): 46-02, 46A05, 46A07, 46A09, 46A15, 46A30

ISBN 3-540-08662-5 Springer-Verlag Berlin Heidelberg New York
ISBN 0-387-08662-5 Springer-Verlag New York Heidelberg Berlin

Printed in Germany

Printing and binding: Beltz Offsetdruck, Hemsbach/Bergstr.
2141/3140-543210

Unserem verehrten Lehrer

Gottfried Köthe

in Dankbarkeit gewidmet

Contents

Introduction

One of the earliest and most important results in functional analysis are the closed graph theorem and the open mapping theorem of Banach and Schauder. Usually these theorems are known for Banach spaces, but already the original version gives their validity for complete metrizable topological vector spaces, a class of spaces including not only the Banach spaces, but also other important spaces for which in general duality methods cannot be applied in the proofs.

But this "non convex" approach was dropped in the following development of functional analysis. Starting from Banach spaces the theory turned to the more general class of locally convex spaces, making extensive use of the close interrelation between the spaces and their dual spaces. This method was very fruitful and let to far reaching results.

A comprehensive presentation of the theory of locally convex spaces can be found in G. KÖTHE: Topologische lineare Räume I . A continuation of considerations in this book and newest results will give the second volume (which probably will appear soon). In this connection we should also mention the books of N. BOURBAKI [1], A. GROTHENDIECK [2], J. KELLEY, I. NAMIOKA e.a. [1], A.P. and W. ROBERTSON [2], H.H. SCHAEFER [1]. However, we point out that we follow G. KÖTHE [4] with regard to the notions in topological vector spaces not defined in the following.

Nevertheless, besides the above mentioned development there were made a few attempts to build up a basic theory of topological vector spaces, for instance in W. ROBERTSON [1]. One of the aims was to find classes of spaces in the category of topological vector spaces, for which well known principles of functional analysis hold true. In this sense W. ROBERTSON succeeded in defining "barrelled" topological vector spaces, a class for which (among other things) a "Banach-Steinhaus theorem" can be proved.

The ideas of W. ROBERTSON were taken up by S.O. IYAHEN [1]. Guided by the locally convex theory he found a satisfactory approach to notions as "barrelled" and "bornological" within the framework of topological vector spaces. For this he had to substitute the absolutely convex sets M and their property $M+M = 2M$ by decreasing sequences (U_n) of absorbing sets U_n with $U_{n+1} + U_{n+1} \subset U_n$, the "defining sequences". In our work these "defining sequences" are denoted by "strings". In this connection, however, we must also mention the papers of J. KÖHN [1], S. TOMASEK [1], [2] and L. WAELBROECK [1].

Starting from the notion of a "string" in a vector space we develop a general theory of topological vector spaces giving most of the results known up to now (for questions and considerations in this field arising from applications see P. TURPIN [2]). The importance of the strings can be seen from the very beginning. They help to develop a theory of topological vector spaces which gives a satisfactory generalization of the locally convex theory. Of course, in this general situation there is no sufficiently large topological dual space. Nevertheless, there is a surprising validity of many important "locally convex" theorems. To see this one has to find new ways and proofs, proofs which are free from duality theory. It was surprising again (or not at all, as can be said afterwards!), how "elementary and natural" these proofs often became by the necessity to forget about the dual space. Of course, these new proofs can also be used to prove the known results and even new ones in the locally convex theory.

The first 5 sections deliver the general setting of the theory (topological vector spaces, metrizability, projective and inductive limits, topological direct sums).

In sections 6 - 10 we investigate the class of "barrelled" topological vector spaces which is important also in this general theory. The main part of these sections is taken by theorems on linear mappings (the Banach-Steinhaus theorem, closed graph theorems, open mapping theorems).

Section 11 introduces the "bornological" spaces, and in section 12 we deal with spaces of linear mappings and their topologies. Among other things there is given a simple representation of the completion of these spaces.

Interesting generalizations of the class of (DF)-spaces are given in sections 15 - 17 by considering the following property: A subset, which is "large enough", is a neighbourhood of 0, if and only if it induces a neighbourhood of 0 on all bounded balanced sets. Finally, section 18 interprets and completes the foregoing considerations for (DF)-spaces.

We remark that the results in sections 8 - 10 , 12 , 15 - 17 give new results also for the locally convex theory, which were obtained just in the last years.

Hint: If we say "barrelled, bornological, ... ", then this is an abbreviation for the (in the following defined) new notions " $\mathfrak{L}$-barrelled, $\mathfrak{L}$-bornological, ... " (or "barrelled in $\mathfrak{L}$, bornological in $\mathfrak{L}$, ... ", where $\mathfrak{L}$ is the category of topological vector spaces). These are not the notions used in G. KÖTHE [4], for which we say " $\mathfrak{C}$-barrelled, ... " (or "barrelled in $\mathfrak{C}$, ... " with $\mathfrak{C}$ as the category of locally convex spaces).

N. Adasch and D. Keim decided and began to write down this systematic development of the theory of topological vector spaces. From their close collaboration resulted N. Adasch's lecture: "Nicht lokalkonvexe Vektorräume" in 1972/1973 , which became a basic part of this monography. Since they wanted to consider also the theory of (DF)-spaces (developped in B. ERNST [1] and J.P. LIGAUD [1]) and, more generally, of locally topological spaces (N. ADASCH [7], N. ADASCH, B. ERNST [3], [4]), it was quite natural that B. Ernst joined to write these notes.

§ 1 Strings and linear topologies

In this and the following chapters we shall only consider vector spaces over the field $\mathbb{K}$ of real or complex numbers, $\mathbb{K}$ equipped with the usual topology. For all the notions of the theory of (topological) vector spaces which we do not define here, see G. KÖTHE [4].

Let E be a vector space over $\mathbb{K}$. A sequence $\mathfrak{U} = (U_n)$ of subsets U_n of E is called a s t r i n g (in E) (in German: Faden), if

(i) every $U_n \in \mathfrak{U}$ is b a l a n c e d , that means for any $x \in U_n$ and $\lambda \in \mathbb{K}$, $|\lambda| \leq 1$, we have $\lambda x \in U_n$,

(ii) every U_n is a b s o r b i n g , that means for any $x \in E$ there is a $\lambda \in \mathbb{K}$, $\lambda > 0$, such that $x \in \lambda U_n$,

(iii) (U_n) is s u m m a t i v e , that means $U_{n+1} + U_{n+1} \subset U_n$ for all $n \in \mathbb{N}$.

U_1 is called the b e g i n n i n g of the string $\mathfrak{U}$, and U_n is the nth k n o t of $\mathfrak{U}$.

If $\mathfrak{U} = (U_n)$ and $\mathfrak{V} = (V_n)$ are strings in E and if $\lambda \in \mathbb{K}$, we define

$$\lambda\mathfrak{U} := (\lambda U_n) ,$$
$$\mathfrak{U} + \mathfrak{V} := (U_n + V_n) ,$$
$$\mathfrak{U} \cap \mathfrak{V} := (U_n \cap V_n) .$$

$\mathfrak{U} + \mathfrak{V}$ is called the s u m and $\mathfrak{U} \cap \mathfrak{V}$ the i n t e r s e c t i o n of the strings $\mathfrak{U}$ and $\mathfrak{V}$. Of course $\mathfrak{U} + \mathfrak{V}$, $\mathfrak{U} \cap \mathfrak{V}$ and $\lambda\mathfrak{U}$, $\lambda \neq 0$, are again strings in E.

For a string $\mathfrak{U}$ we denote by $N(\mathfrak{U})$ the set $N(\mathfrak{U}) := \bigcap_{n \in \mathbb{N}} U_n$. $N(\mathfrak{U})$ is called the k e r n e l of $\mathfrak{U}$. Since a string $\mathfrak{U}$ is summative and its knots are balanced, the kernel $N(\mathfrak{U})$ is a linear subspace of E.

We write $\mathfrak{U} \subset \mathfrak{V}$ for two strings $\mathfrak{U} = (U_n)$ and $\mathfrak{V} = (V_n)$, if this relation holds for the knots, i.e. if $U_n \subset V_n$ for all $n \in \mathbb{N}$.

If the vector space E is equipped with a topology $\mathfrak{T}$, we denote this topological space by $E(\mathfrak{T})$. $\mathfrak{T}$ is a *linear topology*, if addition and scalar multiplication are continuous mappings from $E(\mathfrak{T}) \times E(\mathfrak{T})$ and $\mathbb{K} \times E(\mathfrak{T})$ into $E(\mathfrak{T})$. If $\mathfrak{T}$ is linear and Hausdorff, we call $E(\mathfrak{T})$ a *topological vector space* (abbreviated: t.v.s.).

Let $\mathfrak{T}$ be a linear topology on E. A string $\mathfrak{U} = (U_n)$ in $E(\mathfrak{T})$ is called a ($\mathfrak{T}$-) *topological string*, if every knot U_n is a $\mathfrak{T}$-neighbourhood of O. A neighbourhood U of O in $E(\mathfrak{T})$ generates a (not uniquely determined) topological string: U contains a balanced O-neighbourhood U_1, U_1 contains a balanced O-neighbourhood U_2 with $U_2 + U_2 \subset U_1$, ...

The connection between strings and linear topologies gives the proposition

(1) <u>Let $\mathfrak{T}$ be a linear topology on the vector space E. Then there is a set $\mathfrak{F}$ of strings in E with the following properties:</u>

(i) <u>If $\mathfrak{U} \in \mathfrak{F}$ and $\mathfrak{V} \in \mathfrak{F}$, there is a $\mathfrak{W} \in \mathfrak{F}$, such that $\mathfrak{W} \subset \mathfrak{U} \cap \mathfrak{V}$.</u>

(ii) <u>The knots of the strings in $\mathfrak{F}$ form a base of O-neighbourhoods in $E(\mathfrak{T})$.</u>

<u>For such a set $\mathfrak{F}$ we have: $\mathfrak{T}$ is Hausdorff, if and only if $\bigcap_{\mathfrak{U} \in \mathfrak{F}} N(\mathfrak{U}) = \{0\}$.</u>

For instance, every base of O-neighbourhoods in $E(\mathfrak{T})$ gives us in an obvious way a set $\mathfrak{F}$ of strings with (i) and (ii). In particular the set of all topological strings in $E(\mathfrak{T})$ satisfies conditions (i) and (ii).

Conversely, how certain strings generate a linear topology is shown by

(2) *Let $\mathfrak{F}$ be a set of strings in a vector space E, such that*

(i) *for all $\mathfrak{U}, \mathfrak{V} \in \mathfrak{F}$ there is a $\mathfrak{W} \in \mathfrak{F}$ with $\mathfrak{W} \subset \mathfrak{U} \cap \mathfrak{V}$.*

Then the knots of the strings in $\mathfrak{F}$ form a base of 0-neighbourhoods for a linear topology $\mathfrak{T}_{\mathfrak{F}}$ on E.

If furthermore

$$\text{(ii)} \qquad \bigcap_{\mathfrak{U} \in \mathfrak{F}} N(\mathfrak{U}) = \{0\},$$

then $\mathfrak{T}_{\mathfrak{F}}$ is Hausdorff and $E(\mathfrak{T}_{\mathfrak{F}})$ is a t.v.s..

A set $\mathfrak{F}$ of strings in E with property (2), (i) is called **directed**. A directed set $\mathfrak{F}$ of strings in a t.v.s. $E(\mathfrak{T})$ with $\mathfrak{T} = \mathfrak{T}_{\mathfrak{F}}$ is called ($\mathfrak{T}$-) **fundamental**, i.e. the knots of the strings in $\mathfrak{F}$ form a base of 0-neighbourhoods in $E(\mathfrak{T})$. In this case we say: $\mathfrak{F}$ **generates** $\mathfrak{T}$.

(2) follows from G. KÖTHE [4], § 15, 2.(2): the knots of the strings in $\mathfrak{F}$ form a filter base generating a linear topology on E.

The set of all strings in a vector space E generates by (2) a linear topology $\mathfrak{T}^f$ on E. By (1) $\mathfrak{T}^f$ is the finest linear topology on E.

Every absorbing absolutely convex set U in E gives a string $\mathfrak{U}_U = (U_n)$ with $U_n := \frac{1}{2^{n-1}} U$. $\mathfrak{U}_U$ is called the **natural string of U**.

A linear topology $\mathfrak{T}$ on E is called **locally convex**, if $\mathfrak{T}$ has a base of 0-neighbourhoods of absolutely convex sets.

The set of the natural strings of all absorbing absolutely convex sets in E generates by (2) a locally convex topology $\mathfrak{T}^c$ on E. $\mathfrak{T}^c$ is the finest locally convex topology on E and $\mathfrak{T}^c \subset \mathfrak{T}^f$.

Now we will consider some special strings in the vector space E. Let $\{e_\alpha\}_{\alpha\in I}$ be an algebraic basis of E. For a positive real number p we define the string $\mathfrak{U}^p = (U_n^p)$ by

$$U_n^p := \{x\colon x = \sum_{\alpha\in I}\lambda_\alpha e_\alpha \text{ with } \sum_{\alpha\in I}|\lambda_\alpha|^p \le \tfrac{1}{2^{n-1}}\},\ \text{if } 0<p\le 1,\quad {}^{*)}$$

$$U_n^p := \{x\colon x = \sum_{\alpha\in I}\lambda_\alpha e_\alpha \text{ with } (\sum_{\alpha\in I}|\lambda_\alpha|^p)^{\frac{1}{p}} \le \tfrac{1}{2^{n-1}}\},\ \text{if } 1\le p.$$

For all p we have $N(\mathfrak{U}^p) = \{0\}$. For $p \ge 1$ the string $\mathfrak{U}^p$ is absolutely convex, i.e. all U_n^p are absolutely convex. Hence

(3) $\mathcal{T}^c$ and $\mathcal{T}^f$ are Hausdorff linear topologies on E.

For $p \ge 1$ all U_n^p are absolutely convex. For $0 < p < 1$ we have another situation.

(4) For p with $0 < p < 1$ the following assertions on E are equivalent:

(i) E has an uncountable algebraic dimension dim E.

(ii) The beginning U_1^p of $\mathfrak{U}^p$ contains no absorbing absolutely convex set.

Proof. (i) ⇒ (ii): Let $\{e_\alpha\}_{\alpha\in I}$ be a basis in E. If there is an absorbing absolutely convex set U in E with $U \subset U_1^p$, there exist $\varepsilon_\alpha > 0$ with $\varepsilon_\alpha e_\alpha \in U$ for all $\alpha \in I$. Since I is uncountable, we can find $\varepsilon > 0$, such that $I_\varepsilon := \{\alpha\in I\colon \varepsilon_\alpha \ge \varepsilon\}$ is infinite. If $\alpha_1,\dots,\alpha_n \in I_\varepsilon$, we have $\frac{\varepsilon}{\varepsilon_{\alpha_l}}\varepsilon_{\alpha_l}e_{\alpha_l} \in U$ and $\sum_{l=1}^{n}\frac{\varepsilon}{n}e_{\alpha_l} \in U$. On the other hand $U \subset U_1^p$ and therefore

$$1 \ge \sum_{1}^{n}\left(\frac{\varepsilon}{n}\right)^p = n^{1-p}\varepsilon^p .$$

Since n was arbitrary, this is a contradiction.

---

*) We consider sums $\sum_{\alpha\in I}\lambda_\alpha e_\alpha$, in which $\lambda_\alpha \neq 0$ for at most finitely many α.

"(ii) ⇒ (i)" is a consequence of

(5) If dim E is countable, then every string $\mathfrak{U} = (U_n)$ in E contains an absolutely convex string.

Proof. Let $\{e_n\}_{n\in\mathbb{N}}$ be an algebraic basis of E. If we choose $\varepsilon_\ell^{(1)} > 0$ with $\varepsilon_\ell^{(1)} e_\ell \in U_{1+\ell}$ and $\varepsilon_\ell^{(n)}$ with $0 < \varepsilon_\ell^{(n)} \leq \frac{1}{2}\varepsilon_\ell^{(n-1)}$ and $\varepsilon_\ell^{(n)} e_\ell \in U_{n+\ell}$ for $n > 1$, then the $V_n := \sum_{\ell=1}^{\infty} \varepsilon_\ell^{(n)}[e_\ell]_1$ *) form an absolutely convex string $\mathfrak{V} = (V_n)$ in E. We have $\mathfrak{V} \subset \mathfrak{U}$ because of

$$V_n = \sum_{\ell=1}^{\infty} \varepsilon_\ell^{(n)}[e_\ell]_1 \subset \sum_{\ell=1}^{\infty} U_{n+\ell} \subset U_n .$$

We collect the last results in (see J.L. KELLEY, I. NAMIOKA e.a. [1], J. KÖHN [1])

(6) If dim E is countable, then $\mathcal{T}^f = \mathcal{T}^c$. If dim E is uncountable, then $\mathcal{T}^f$ is strictly finer than $\mathcal{T}^c$.

---

*) For an index set I and $M_\alpha \subset E$, $\alpha \in I$, we define $\sum_{\alpha\in I} M_\alpha := \bigcup_{e\subset I} (\sum_{\alpha\in e} M_\alpha)$, where e is a finite subset of I.
If $x \in E$, $\lambda > 0$, then $[x]_\lambda := \{\mu x : |\mu| \leq \lambda\}$.

§ 2 Metrizable topological vector spaces

In this section we consider a first and very important class of t.v.s.. We call a t.v.s. $E(\mathcal{T})$ m e t r i z a b l e , if there is a metric ϱ on E, such that for all $x \in E$ the balls $K_\varepsilon(x) := \{y: y \in E \text{ with } \varrho(x,y) < \varepsilon\}$ (or equivalently the balls $\bar{K}_\varepsilon(x) := \{y: \varrho(x,y) \le \varepsilon\}$) form a base of $\mathcal{T}$-neighbourhoods of x. Then $\{K_{\frac{1}{n}}(0)\}_{n \in \mathbb{N}}$ is a base of 0-neighbourhoods in $E(\mathcal{T})$, and we obtain:

(1) <u>If</u> $E(\mathcal{T})$ <u>is metrizable, there is a topological string</u> $\mathfrak{U} = (U_n)$ <u>in</u> $E(\mathcal{T})$, <u>whose knots</u> U_n <u>form a base of 0-neighbourhoods, i.e. an</u> $\mathfrak{U}$ <u>with</u> $\mathcal{T} = \mathcal{T}_{\mathfrak{U}}$.*)

For the proof choose a topological string $\mathfrak{U} = (U_n)$ with $U_n \subset K_{\frac{1}{n}}(0)$.

Conversely we want to show that a t.v.s. $E(\mathcal{T})$ is metrizable, if there is a string $\mathfrak{U}$ in E with $\mathcal{T} = \mathcal{T}_{\mathfrak{U}}$. For this we need the notion of an (F)-norm.

An (F) - n o r m on a vector space E is a real valued function $\| \; \|: E \to \mathbb{R}$, such that for all $x,y \in E$ we have

(F 1) $\|x\| \ge 0$,

(F 2) $\|x\| = 0 \Rightarrow x = 0$,

(F 3) $\|x+y\| \le \|x\| + \|y\|$,

(F 4) $\|\lambda x\| \le \|x\|$ for $\lambda \in \mathbb{K}$ with $|\lambda| \le 1$,

(F 5) if $\lambda_n \to 0$, $\lambda_n \in \mathbb{K}$, then $\|\lambda_n x\| \to 0$.

There are many (F)-norms on a vector space E. For instance, if for $x \in E$ we have the representation $x = \sum_{\alpha \in I} \lambda_\alpha e_\alpha$, where $\{e_\alpha\}$ is a basis in E, then define

---

*) $\mathcal{T}_{\mathfrak{U}} := \mathcal{T}_{\mathfrak{F}}$ with $\mathfrak{F} = \{\mathfrak{U}\}$, see 1.(2).

$$\|x\|_p := \sum_{\alpha\in I} |\lambda_\alpha|^p \text{ , if } 0 < p \leq 1,$$
$$\|x\|_p := \left(\sum_{\alpha\in I} |\lambda_\alpha|^p\right)^{\frac{1}{p}} \text{ , if } p \geq 1.$$

Each $\| \ \|_p$ is an (F)-norm on E (and no norm, if $0 < p < 1$).

An (F)-norm $\| \ \|$ generates a metric ϱ on E by $\varrho(x,y) := \|x-y\|$. This metric ϱ is translation invariant, i.e. $\varrho(x,y) = \varrho(x+z,y+z)$ for all $x,y,z \in E$. The sets $K_{\frac{1}{2^{n+1}}}(0)$ form a string $\mathfrak{U}$ in E with $N(\mathfrak{U}) = \{0\}$. The topology $\mathcal{T}_{\mathfrak{U}}$ generated by $\mathfrak{U}$ coincides with the topology generated by the metric ϱ. This follows from $K_\varepsilon(x) = x + K_\varepsilon(0)$.

Let conversely $\mathfrak{U} = (U_n)$ be a string in E, such that $N(\mathfrak{U}) = \{0\}$. We want to construct an (F)-norm on E, such that $\mathcal{T}_{\mathfrak{U}}$ and the topology generated by the (F)-norm coincide.

Consider the rational dyadic numbers $\delta > 0$ of the form ($n = 0$ or $n \in \mathbb{N}$)

$$\delta = n + \sum_{k=1}^{\infty} \varepsilon_k \frac{1}{2^k}, \text{ where}$$
$$\varepsilon_k = \begin{cases} 1 & \text{for at most finitely many } k \in \mathbb{N}, \\ 0 & \text{otherwise.} \end{cases}$$

For every δ we define a set W_δ by

$$W_\delta := \sum_1^n U_1 + \sum_{k=1}^{\infty} \varepsilon_k U_{k+1} .$$

Then for $x \in E$ we set

$$\|x\| := \inf\{\delta : x \in W_\delta\} .$$

We want to show, that $\| \ \|$ is an (F)-norm on the vector space E.

<u>(F 1)</u> is clear. <u>(F 2)</u>: Take $x \in E$ with $\|x\| = 0$. Then we have $x \in W_\delta$ for all δ, especially $x \in U_k$ for all $k \in \mathbb{N}$. Hence $x \in N(\mathfrak{U}) = \{0\}$.

<u>(F 3)</u> For two numbers δ_1 and δ_2 we have $W_{\delta_1} + W_{\delta_2} \subset W_{\delta_1+\delta_2}$, hence $x \in W_{\delta_1}$ and $y \in W_{\delta_2}$ implies $\|x+y\| \leq \delta_1 + \delta_2$. From this follows (F 3).

<u>(F 4)</u> is obvious, since all W_δ are balanced.

<u>(F 5)</u>: Since all U_n are absorbing and balanced, there is for a given $k \in \mathbb{N}$ an $n_0 \in \mathbb{N}$, such that $\lambda_n x \in U_{k+1}$ for $n > n_0$. That means $\|\lambda_n x\| \leq \frac{1}{2^k}$ for $n > n_0$.

Thus we have shown that every string $\mathfrak{U}$ in E with $N(\mathfrak{U}) = \{0\}$ generates an (F)-norm $\| \ \|_{\mathfrak{U}}$ and hence a translation invariant metric $\varrho_{\mathfrak{U}}$ on E.

If $\mathfrak{U}$ is the natural string of an absorbing absolutely convex set U in E, then $W_\delta = \delta U$ holds and therefore $\|x\|_{\mathfrak{U}} = q_U(x)$ for all $x \in E$, where q_U denotes the Minkowski functional of U.

(2) Let $\mathfrak{U} = (U_n)$ be a string, such that $N(\mathfrak{U}) = \{0\}$. Then $\mathfrak{T}_{\mathfrak{U}}$ coincides with the topology generated by $\| \ \|_{\mathfrak{U}}$ resp. $\varrho_{\mathfrak{U}}$, and $K_{\frac{1}{2^n}}(0) \subset U_{n+1} \subset K_{\overline{\frac{1}{2^n}}}(0)$ holds.

Proof. It suffices to show the last relation. If $x \in K_{\frac{1}{2^n}}(0)$, i.e. $\|x\|_{\mathfrak{U}} < \frac{1}{2^n}$, there is a $\delta = \sum_{k=1}^{\infty} \varepsilon_k \frac{1}{2^k}$ with $\varepsilon_k = 0$ for $1 \leq k \leq n$, such that $x \in W_\delta = \sum_{k=n+1}^{\infty} \varepsilon_k U_{k+1} \subset U_{n+1}$. For $x \in U_{n+1}$ holds $\|x\|_{\mathfrak{U}} \leq \frac{1}{2^n}$ and $x \in K_{\overline{\frac{1}{2^n}}}(0)$.

It is easy to see that we have $\overline{U_{n+1}}^{\mathfrak{T}_{\mathfrak{U}}} = K_{\overline{\frac{1}{2^n}}}(0)$ in (2), but $\overline{K_{\frac{1}{2^n}}(0)}^{\mathfrak{T}_{\mathfrak{U}}} \neq K_{\overline{\frac{1}{2^n}}}(0)$ in general. Combining (1) and (2) we obtain

(3) The t.v.s. $E(\mathfrak{T})$ is metrizable, if and only if there is a string $\mathfrak{U}$ in E, whose knots are a base of neighbourhoods of 0 in $E(\mathfrak{T})$, i.e. for which $\mathfrak{T}_{\mathfrak{U}} = \mathfrak{T}$.

Especially there exists on every metrizable t.v.s. a translation invariant metric generating the topology.

The problem of metrizability for general topological spaces was unsolved until 1950. Our simpler question of metrizability of a t.v. s. was earlier answered by S. KAKUTANI [1] in a way similar as done here (for the construction of an (F)-norm from a string $\mathfrak{U}$ see S. ROLEWICZ [1], L. WAELBROECK [1]).

A metrizable complete t.v.s. is called (F) - s p a c e .

If $\mathfrak{U}$ is a string in a vector space E with $N(\mathfrak{U}) = \{0\}$, then $E(\mathcal{T}_{\mathfrak{U}})$ is metrizable by (2), and the completion $\widetilde{E}(\widetilde{\mathcal{T}_{\mathfrak{U}}})$ of $E(\mathcal{T}_{\mathfrak{U}})$ is an (F)-space. If $\mathfrak{U}$ is an arbitrary string, we consider the quotient space $E/N(\mathfrak{U})$. Denote by $K_{\mathfrak{U}}$ the canonical mapping from E onto $E/N(\mathfrak{U})$. Then $K_{\mathfrak{U}}(\mathfrak{U}) := (K_{\mathfrak{U}}(U_n))$ is a string in $E/N(\mathfrak{U})$, and we have $N(K_{\mathfrak{U}}(\mathfrak{U})) = \{0\}$. Hence $E/N(\mathfrak{U})(\mathcal{T}_{K_{\mathfrak{U}}(\mathfrak{U})})$ is a metrizable t.v.s.. This space is denoted by $E_{\mathfrak{U}}$ and its completion by $\widetilde{E_{\mathfrak{U}}}$. $K_{\mathfrak{U}}$ again denotes the canonical mapping from E into $\widetilde{E_{\mathfrak{U}}}$.

If $E(\mathcal{T})$ is a t.v.s. and $\mathfrak{U}$ a string in E, then are equivalent: (i) $\mathfrak{U}$ is a topological string in $E(\mathcal{T})$, (ii) $K_{\mathfrak{U}}: E(\mathcal{T}) \longrightarrow E_{\mathfrak{U}}$ is continuous, (iii) $K_{\mathfrak{U}}: E(\mathcal{T}) \longrightarrow \widetilde{E_{\mathfrak{U}}}$ is continuous. (i) $\Longrightarrow$ (ii) $\Longrightarrow$ (iii) is clear, (iii) $\longrightarrow$ (i): If $K_{\mathfrak{U}}$ is continuous, then $K_{\mathfrak{U}}^{-1}(\widetilde{K_{\mathfrak{U}}(U_{n+2})})$ is a neighbourhood of 0 in $E(\mathcal{T})$ ($\widetilde{K_{\mathfrak{U}}(U_{n+2})}$ is the closure of $K_{\mathfrak{U}}(U_{n+2})$ in $\widetilde{E_{\mathfrak{U}}}$). But $K_{\mathfrak{U}}^{-1}(\widetilde{K_{\mathfrak{U}}(U_{n+2})}) = K_{\mathfrak{U}}^{-1}(\widetilde{K_{\mathfrak{U}}(U_{n+2})} \cap E_{\mathfrak{U}}) \subset K_{\mathfrak{U}}^{-1}(K_{\mathfrak{U}}(U_{n+2}) + K_{\mathfrak{U}}(U_{n+2})) = U_{n+2} + U_{n+2} + N(\mathfrak{U}) \subset U_n$, hence U_n is a neighbourhood of 0 in $E(\mathcal{T})$.

Examples.

1) The most important examples of metrizable topological vector spaces are the metrizable locally convex spaces and among these especially the Banach and Hilbert spaces.

2) Let Σ be a σ-algebra of subsets of a set X and μ a measure on Σ. Let $\varphi(t)$ for $t \geq 0$ be a continuous, increasing, nonnegative function with: $\varphi(t) = 0$ if and only if $t = 0$, $\varphi(2t) \leq k\,\varphi(t)$ for a certain k and all t. We denote by

$$L^{\varphi}(X, \Sigma, \mu)$$

the linear space of all μ-equivalence classes f of measurable functions on X with

$$\int_X \varphi(|f(t)|)\, d\mu < \infty .$$

For certain $\varepsilon_n > 0$ the sets

$$U_n := \left\{ f: \int_X \varphi(|f(t)|)\, d\mu < \varepsilon_n \right\}$$

give us a string in $L^{\varphi}(X,\Sigma,\mu)$ and generate a metrizable linear topology on these spaces. Moreover the $L^{\varphi}(X,\Sigma,\mu)$ are (F)-spaces with these topologies.

Choosing X, Σ, μ in a concreter form one obtains many well known (F)-spaces:

a) If $\varphi(t) = t^p$, $0 < p < \infty$, $X = [0,1]$ the unit intervall, Σ the Lebesgue measurable subsets and μ the Lebesgue measure, then

$$L^{\varphi}(X,\Sigma,\mu) = L^p(0,1) .$$

This space is a Banach space for $1 \le p < \infty$, hence locally convex. For $0 < p < 1$ it is not locally convex (see G. KÖTHE [4], §15, 9.).

b) With $\varphi(t) = t^p$, $0 < p < \infty$, and $X = \mathbb{N}$, Σ the set of all subsets of $\mathbb{N}$ and μ the counting measure on Σ we get with

$$L^{\varphi}(X,\Sigma,\mu) = \ell^p$$

the usual sequence spaces.

c) If $\varphi(t) = \frac{t}{1+t}$, $X = [0,1]$, Σ all Lebesgue measurable subsets and μ the Lebesgue measure, then we have with

$$L^{\varphi}(X,\Sigma,\mu) = S(0,1)$$

the space of all measurable functions on $[0,1]$ endowed with the topology of "convergence in measure".

3) "Mappings with rapidly decreasing approximation numbers": Let E be a Banach space and L(E) the space of all continuous operators on E endowed with the operator norm. For $A \in L(E)$ one defines the kth approximation number by

$$\alpha_k(A) := \inf \left\{ \|B - A\| : B \in L(E) \text{ with } \dim B(E) \le k \right\}.$$

If

$$S(E) := \left\{ A: A \in L(E) \text{ with } \sum_{k=0}^{\infty} (\alpha_k(A))^p < \infty \text{ for all } p > 0 \right\},$$

then on S(E) a metrizable (not locally convex) linear topology is generated by the string (with appropriate $\varepsilon_n > 0$, $p_n > 0$)

$$U_n: \; = \left\{A: \; \sum_{k=0}^{\infty} (\alpha_k(A))^{p_n} < \varepsilon_n\right\}.$$

With this topology S(E) is an (F)-space (see A. PIETSCH [1], chapter 8, B. GRAMSCH [1]).

§ 3 Projective limits of topological vector spaces

In this section the importance of (F)-spaces for the theory of topological vector spaces will become clear. It will be shown, that every t.v.s. can be generated by (F)-spaces in a certain sense.

Let $E(\mathfrak{T})$ be a t.v.s. and let $\mathfrak{F}$ be a set of strings in E. $\mathfrak{F}$ is called ($\mathfrak{T}$-) *saturated*, if all the elements of $\mathfrak{F}$ are topological and if for any $\mathfrak{T}$-topological string $\mathfrak{W}$ in $E(\mathfrak{T})$ there is an $\mathfrak{U} \in \mathfrak{F}$ with $\mathfrak{U} \subset \mathfrak{W}$. The knots of the strings of such a saturated $\mathfrak{F}$ form a base of neighbourhoods of 0 in $E(\mathfrak{T})$.

(1) Let E be a vector space and I an index set. Suppose for all $\alpha \in I$ is given a t.v.s. $E_\alpha(\mathfrak{T}_\alpha)$ and a linear mapping A_α from E into $E_\alpha(\mathfrak{T}_\alpha)$. Suppose that in every $E_\alpha(\mathfrak{T}_\alpha)$ a saturated set $\mathfrak{F}_\alpha$ of strings is given. For finite subsets $\{\alpha_1, \dots, \alpha_m\}$ in I and for $\mathfrak{U}_{\alpha_i} \in \mathfrak{F}_{\alpha_i}$ we consider the strings

$$\bigcap_{i=1}^{m} A_{\alpha_i}^{-1}(\mathfrak{U}_{\alpha_i}) \ . \quad {}^{*)}$$

Let $\mathfrak{F}$ denote the set of all such strings in E. Then we have:

(i) $\mathfrak{F}$ is directed.

(ii) The linear topology $\mathfrak{T}_{\mathfrak{F}}$ is the coarsest topology on E, for which the mappings A_α, $\alpha \in I$, are continuous.

(iii) $\mathfrak{T}_{\mathfrak{F}}$ is Hausdorff, if and only if $\bigcap_{\alpha \in I} N(A_\alpha) = \{0\}$.

We can omit the simple proof of (1). $\mathfrak{T}_{\mathfrak{F}}$ is called the *projective topology* on E with respect to $\{E_\alpha(\mathfrak{T}_\alpha), A_\alpha\}_{\alpha \in I}$.

---

*) $A_\alpha^{-1}(\mathfrak{U}_\alpha)$ denotes the string $(A_\alpha^{-1}(U_n^\alpha))$, if $\mathfrak{U}_\alpha = (U_n^\alpha)$.

A t.v.s. $E(\mathfrak{T})$, generated as described in (1), is called the **projective limit** of the spaces $E_\alpha(\mathfrak{T}_\alpha)$ with respect to the linear mappings A_α, and we denote it by $E(\mathfrak{T}) = \operatorname{proj}_{\alpha\in I}(E_\alpha(\mathfrak{T}_\alpha), A_\alpha)$ (here our terminology is different from G. KÖTHE [4]).

Examples. (i) Let H be a subspace of the t.v.s. $E(\mathfrak{T})$ and let $\hat{\mathfrak{T}}$ be the topology on H induced by $\mathfrak{T}$. Then we have $H(\hat{\mathfrak{T}}) = \operatorname{proj}_H(E(\mathfrak{T}), I_H)$, where I_H denotes the embedding of H into E.

(ii) Let $E(\mathfrak{T}) := \prod_{\alpha\in I} E_\alpha(\mathfrak{T}_\alpha)$ be the topological product of the t.v.s. $E_\alpha(\mathfrak{T}_\alpha)$, $\alpha \in I$, and let P_α be the projections from $\prod_{\alpha\in I} E_\alpha$ onto E_α. Then $E(\mathfrak{T}) = \operatorname{proj}_{\alpha\in I}(E_\alpha(\mathfrak{T}_\alpha), P_\alpha)$.

(iii) Suppose it is given a set of Hausdorff linear topologies $\mathfrak{T}_\alpha$, $\alpha \in I$, on a vector space E. If $\mathfrak{T} := \sup_{\alpha\in I} \mathfrak{T}_\alpha$, then $E(\mathfrak{T}) = \operatorname{proj}_{\alpha\in I}(E(\mathfrak{T}_\alpha), I_\alpha)$, where I_α is the identity mapping on E for all α.

The following proposition is very useful and its proof is obvious.

(2) If $E(\mathfrak{T}) = \operatorname{proj}_{\alpha\in I}(E_\alpha(\mathfrak{T}_\alpha), A_\alpha)$ and if $F(\mathfrak{T}')$ is an arbitrary topological space, then a mapping $A: F(\mathfrak{T}') \longrightarrow E(\mathfrak{T})$ is continuous, if and only if all the mappings $A_\alpha \circ A$, $\alpha \in I$, are continuous.

Topological products and their subspaces give the class of all projective limits:

(3) Assume $E(\mathfrak{T}) = \operatorname{proj}_{\alpha\in I}(E_\alpha(\mathfrak{T}_\alpha), A_\alpha)$. Then the mapping J from $E(\mathfrak{T})$ into $\prod_{\alpha\in I} E_\alpha(\mathfrak{T}_\alpha)$ defined by $J(x) := (A_\alpha x)_{\alpha \in I}$ gives a topological isomorphism between $E(\mathfrak{T})$ and a subspace of the product.

Proof. By (1), (iii) we have $\bigcap_{\alpha\in I} N(A_\alpha) = \{0\}$ and hence J is one-to-one. Continuity of J and J^{-1} is a consequence of (2).

Let $\mathfrak{F}$ be a $\mathfrak{T}$-fundamental set of strings in the t.v.s. $E(\mathfrak{T})$, and let $\mathfrak{T}_P$ be the projective topology on E with respect to $\{E_{\mathfrak{U}}, K_{\mathfrak{U}}\}_{\mathfrak{U}\in\mathfrak{F}}$. Since $K_{\mathfrak{U}}: E(\mathfrak{T}) \longrightarrow E_{\mathfrak{U}}$ is continuous for all $\mathfrak{U} \in \mathfrak{F}$, we have $\mathfrak{T}_P \subset \mathfrak{T}$. On the other hand if U is a $\mathfrak{T}$-neighbourhood of 0, there is a knot U_n of a string $\mathfrak{U} = (U_n) \in \mathfrak{F}$ such that $U_n \subset U$. $K_{\mathfrak{U}}^{-1}(K_{\mathfrak{U}}(\mathfrak{U}))$ is $\mathfrak{T}_P$-topological by (1) and $K_{\mathfrak{U}}^{-1}(K_{\mathfrak{U}}(U_{n+1})) \subset U_{n+1} + N(\mathfrak{U}) \subset U_n \subset U$. Hence U is a $\mathfrak{T}_P$-neighbourhood of 0, and we have shown $\mathfrak{T} = \mathfrak{T}_P$.

Therefore $E(\mathfrak{T}) = \underset{\mathfrak{U}\in\mathfrak{F}}{\text{proj}}(E_{\mathfrak{U}}, K_{\mathfrak{U}})$ and $E(\mathfrak{T}) = \underset{\mathfrak{U}\in\mathfrak{F}}{\text{proj}}(\widetilde{E_{\mathfrak{U}}}, K_{\mathfrak{U}})$, which gives us

(4) Every t.v.s. $E(\mathfrak{T})$ is a projective limit of (F)-spaces.

With (3) and (4) we obtain furthermore

(5) Every t.v.s. $E(\mathfrak{T})$ is topologically isomorphic to a subspace of a product of (F)-spaces.

§ 4 Inductive limits of topological vector spaces

As the projective limit the inductive limit gives another useful and important method to construct a t.v.s. using a given system of spaces.

For the definition of an inductive limit we need

(1) Let E be a vector space. For $\alpha \in I$, I an index set, let $A_\alpha : E_\alpha(\mathcal{T}_\alpha) \longrightarrow E$ be a linear mapping from the t.v.s. $E_\alpha(\mathcal{T}_\alpha)$ into E. Assume $E = \sum_{\alpha \in I} A_\alpha(E_\alpha)$ and consider the set $\mathfrak{F}$ of strings in E given by

$$\mathfrak{F} := \{\mathfrak{U} : \mathfrak{U} \text{ is a string in } E \text{ and } A_\alpha^{-1}(\mathfrak{U}) \text{ is a topological string in } E_\alpha(\mathcal{T}_\alpha) \text{ for every } \alpha \in I\}.$$

Then we have

(i) $\mathfrak{F}$ is directed.

(ii) The topology $\mathcal{T}_{\mathfrak{F}}$ is the finest linear topology on E such that all mappings A_α, $\alpha \in I$, are continuous.

The proof of (1) is obvious. The topology $\mathcal{T}_{\mathfrak{F}}$ is called the *inductive topology* on E with respect to $\{E_\alpha(\mathcal{T}_\alpha), A_\alpha\}_{\alpha \in I}$. A t.v.s. $E(\mathcal{T})$ constructed as in (1) is called the *inductive limit of the spaces* $E_\alpha(\mathcal{T}_\alpha)$ with respect to A_α, and we denote it by $E(\mathcal{T}) = \sum_{\alpha \in I} A_\alpha(E_\alpha(\mathcal{T}_\alpha))$ *). Note that according to our definition we require an inductive limit $E(\mathcal{T})$ always to be Hausdorff.

If $E(\mathcal{T})$ is the inductive limit of $E_\alpha(\mathcal{T}_\alpha)$, we can construct a set of strings generating $\mathcal{T}$ using $\mathcal{T}_\alpha$-saturated sets of strings.

---

*) Cf. the construction of the locally convex hull of locally convex spaces in G. KÖTHE [4].

(2) *If* $E(\mathfrak{T}) = \sum_{\alpha\in I} A_\alpha(E_\alpha(\mathfrak{T}_\alpha))$, *choose for* $\alpha\in I$ *a* $\mathfrak{T}_\alpha$*-saturated set* $\mathfrak{F}_\alpha$ *of strings in* $E_\alpha(\mathfrak{T}_\alpha)$. *If* $\mathfrak{U}_\alpha = (U_n^\alpha) \in \mathfrak{F}_\alpha$, *we construct a string* $\mathfrak{U} = (U_n)$ *in* E *by*

$$U_n := \sum_{k=1}^{\infty}\{\bigcup_{\alpha\in I} A_\alpha(U^\alpha_{2^{n-1}k})\}.$$

The set $\mathfrak{F}$ *of all these strings* $\mathfrak{U}$ *is a saturated set of strings in* $E(\mathfrak{T})$.

Proof. The sequence (U_n) forms a string, since for $n\geq 2$ we have

$$\{\bigcup_{\alpha\in I} A_\alpha(U^\alpha_{2^{n-1}k})\} + \{\bigcup_{\alpha\in I} A_\alpha(U^\alpha_{2^{n-1}k})\} \subset \{\bigcup_{\alpha\in I} A_\alpha(U^\alpha_{2^{n-2}(2k-1)})\} + \{\bigcup_{\alpha\in I} A_\alpha(U_{2^{n-2}2k})\}.$$

The string $\mathfrak{U}$ is also $\mathfrak{T}$-topological: For $\alpha_0\in I$ we have

$$A^{-1}_{\alpha_0}(U_n) = A^{-1}_{\alpha_0}\left(\sum_{k=1}^{\infty}\{\bigcup_{\alpha\in I} A_\alpha(U^\alpha_{2^{n-1}k})\}\right) \supset U^{\alpha_0}_{2^{n-1}k}.$$

Hence $A^{-1}_{\alpha_0}(\mathfrak{U})$ is a topological string in $E_{\alpha_0}(\mathfrak{T}_{\alpha_0})$.

$\mathfrak{F}$ is $\mathfrak{T}$-saturated: Let $\mathfrak{W} = (V_n)$ be a topological string in $E(\mathfrak{T})$. Then $A^{-1}_\alpha(\mathfrak{W})$ is topological in $E_\alpha(\mathfrak{T}_\alpha)$, and there are $\mathfrak{U}_\alpha = (U_n^\alpha) \in \mathfrak{F}_\alpha$, such that $U_n^\alpha \subset A_\alpha^{-1}(V_{n+1})$ for $n\geq 1$ and $\alpha\in I$. Hence $\bigcup_{\alpha\in I} A_\alpha(U_n^\alpha) \subset V_{n+1}$ and

$$U_n = \sum_{k=1}^{\infty}\{\bigcup_{\alpha\in I} A_\alpha(U^\alpha_{2^{n-1}k})\} \subset \sum_{k=1}^{\infty} V_{2^{n-1}k+1} \subset V_n .$$

That means $\mathfrak{U} = (U_n) \subset \mathfrak{W}$.

The situation is not so complicated, if we consider an inductive limit of countably many spaces.

(3) *For* $E(\mathfrak{T}) = \sum_{k=1}^{\infty} A_k(E_k(\mathfrak{T}_k))$ *let* $\mathfrak{F}_k$, $k\in\mathbb{N}$, *be a saturated set of strings in* $E_k(\mathfrak{T}_k)$. *If* $\mathfrak{U}_k = (U_n^k) \in \mathfrak{F}_k$, *we define the string* $\mathfrak{U} = (U_n)$ *in* E *by*

$$U_n := \sum_{k=1}^{\infty} A_k(U_n^k) .$$

The set of all such strings $\mathfrak{U}$ *is saturated in* $E(\mathfrak{T})$.

Proof. It is obvious that $\mathfrak{U}$ is a topological string in $E(\mathfrak{T})$. We have to show that $\{\mathfrak{U}\}$ is saturated: If $\mathfrak{W} = (V_n)$ is a topological string in $E(\mathfrak{T})$, there are strings $\mathfrak{U}_k = (U_n^k) \in \mathfrak{F}_k$, $k \in \mathbb{N}$, such that $U_n^k \subset A_k^{-1}(V_{n+k})$. Hence

$$U_n = \sum_{k=1}^{\infty} A_k(U_n^k) \subset \sum_{k=1}^{\infty} V_{n+k} \subset V_n ,$$

that is $(U_n) \subset \mathfrak{W}$.

The following proposition gives a simple criterion for the continuity of linear mappings from an inductive limit into other t.v.s. (for the proof see (1)).

(4) Let $F(\mathfrak{T}')$ be a t.v.s.. If $E(\mathfrak{T}) = \sum_{\alpha \in I} A_\alpha(E_\alpha(\mathfrak{T}_\alpha))$, a linear mapping $A: E(\mathfrak{T}) \longrightarrow F(\mathfrak{T}')$ is continuous, if and only if $A \circ A_\alpha$ is continuous for each $\alpha \in I$.

Examples. (i) Let H be a closed subspace of the t.v.s. $E(\mathfrak{T})$. For the quotient space $E(\mathfrak{T})/H$ we have $E(\mathfrak{T})/H = \sum_H K_H(E(\mathfrak{T}))$, where $K_H: E \longrightarrow E/H$ is the quotient mapping. (Sometimes we write $E/H(\hat{\mathfrak{T}})$ instead of $E(\mathfrak{T})/H$.)

(ii) For each $\alpha \in I$ let $E_\alpha(\mathfrak{T}_\alpha)$ be a t.v.s. and denote by E the algebraic direct sum of the E_α, $E := \bigoplus_{\alpha \in I} E_\alpha$. Let $\mathfrak{T}$ be the inductive limit topology on E with respect to the embeddings $I_\alpha: E_\alpha(\mathfrak{T}_\alpha) \longrightarrow E$. Since $E \subset \prod_{\alpha \in I} E_\alpha$, and since all $I_\alpha: E_\alpha(\mathfrak{T}_\alpha) \longrightarrow E(\mathfrak{T}_\pi)$ are continuous ($\mathfrak{T}_\pi$ denotes the topology which is induced by $\prod_{\alpha \in I} E_\alpha(\mathfrak{T}_\alpha)$ on E), we have $\mathfrak{T}_\pi \subset \mathfrak{T}$, and the inductive limit topology on E is Hausdorff. $E(\mathfrak{T})$ is called *topological direct sum* *) of the $E_\alpha(\mathfrak{T}_\alpha)$, and we denote this sum by $E(\mathfrak{T}) = \bigoplus_{\alpha \in I} E_\alpha(\mathfrak{T}_\alpha)$.

(iii) Let $\mathfrak{T}_\alpha$, $\alpha \in I$, be a set of Hausdorff linear topologies on a vector space E. Let $I_\alpha: E(\mathfrak{T}_\alpha) \longrightarrow E$ be the identity mapping. Then the inductive topology $\mathfrak{T}$ on E with respect to $\{E(\mathfrak{T}_\alpha), I_\alpha\}_{\alpha \in I}$ is the finest linear topology on E which is coarser than all $\mathfrak{T}_\alpha$. $\mathfrak{T}$ is not necessarily Hausdorff.

---

*) $E(\mathfrak{T})$ is not the topological direct sum in the sense of G. KÖTHE [4], § 18,5.. In general $\mathfrak{T}$ is finer than the sum topology constructed there.

The following proposition shows that we obtain all inductive limits by forming direct sums and quotients.

(5) _If_ $E(\mathfrak{T}) = \sum_{\alpha\in I} A_\alpha(E_\alpha(\mathfrak{T}_\alpha))$ _is an inductive limit,_ $E(\mathfrak{T})$ _is topologically isomorphic to the quotient space_ $\bigoplus_{\alpha\in I} E_\alpha(\mathfrak{T}_\alpha)/N$, _where_ N _is the kernel of the mapping_ $J: \bigoplus_{\alpha\in I} E_\alpha \longrightarrow E$ _with_ $J(x_\alpha) := \sum_{\alpha\in I} A_\alpha x_\alpha$.

Proof. By (4) we can easily show that J is continuous. Hence the mapping $\hat{J}: \bigoplus_{\alpha\in I} E_\alpha(\mathfrak{T}_\alpha)/N \longrightarrow E(\mathfrak{T})$, induced by J, is continuous, and again by (4) $\hat{J}^{-1}$ is continuous.

The projective limit of locally convex spaces is again locally convex by 3.(3). Now we want to consider the same question for inductive limits.

(6) _If_ $E(\mathfrak{T}) = \sum_{k=1}^{\infty} A_k(E_k(\mathfrak{T}_k))$ _is a countable inductive limit of locally convex spaces_ $E_k(\mathfrak{T}_k)$, _then_ $E(\mathfrak{T})$ _is also locally convex._

Since in every $E_k(\mathfrak{T}_k)$ the absolutely convex topological strings form a saturated set of strings, the assertion follows from (3).

For $E(\mathfrak{T}) = \sum_{\alpha\in I} A_\alpha(E_\alpha(\mathfrak{T}_\alpha))$ with I uncountable (6) is in general not true: Let E be a vector space with algebraic basis $\{e_\alpha\}_{\alpha\in I}$, and let $\mathfrak{T}^f$ be the finest linear topology on E. Let $[e_\alpha]$ be the one dimensional subspace generated by e_α, and let $I_\alpha: [e_\alpha] \to E$ be its embedding. Then we have $E(\mathfrak{T}^f) = \sum_{\alpha\in I} I_\alpha([e_\alpha](\mathfrak{T}_\alpha))$, where the natural topology $\mathfrak{T}_\alpha$ of $[e_\alpha]$ is locally convex. But by 1.(6) $E(\mathfrak{T}^f)$ is not locally convex, if dim E is uncountable.

For topological direct sums $E(\mathfrak{T}) = \bigoplus_{\alpha \in I} E_\alpha(\mathfrak{T}_\alpha)$ of locally convex spaces $E_\alpha(\mathfrak{T}_\alpha)$ the converse of (6) is also true: $\bigoplus_{\alpha \in I} E_\alpha(\mathfrak{T}_\alpha)$ is locally convex, if and only if I is countable.

References: S.O. IYAHEN [1], L. WAELBROECK [1].

§ 5 Topological direct sums, strict inductive limits

First we want to prove some further properties of topological direct sums $\bigoplus_{\alpha\in I} E_\alpha(\mathcal{T}_\alpha)$. For a subset I' of I we can embed $\bigoplus_{\alpha\in I'} E_\alpha$ into $\bigoplus_{\alpha\in I} E_\alpha$. The range of this embedding is again denoted by $\bigoplus_{\alpha\in I'} E_\alpha$. Let $\hat{P}_{I'}$ be the projection of $\bigoplus_{\alpha\in I} E_\alpha$ on $\bigoplus_{\alpha\in I'} E_\alpha$. Considering E_α as a subspace of $\bigoplus_{\alpha\in I} E_\alpha$ we obtain the knots of a saturated set of strings of $E(\mathcal{T}) = \bigoplus_{\alpha\in I} E_\alpha(\mathcal{T}_\alpha)$ by (cf. 4.(2))

$$U_n = \sum_{k=1}^{\infty} \{ \bigcup_{\alpha\in I} U^\alpha_{2^{n-1}k} \} . \qquad (*)$$

Now we can show

(1) Let $E_\alpha(\mathcal{T}_\alpha)$, $\alpha \in I$, be a set of t.v.s. and let I' be a subset of I. Then:

(i) $(\bigoplus_{\alpha\in I'} E_\alpha)(\hat{\mathcal{T}}) = \bigoplus_{\alpha\in I'} E_\alpha(\mathcal{T}_\alpha)$, where $\hat{\mathcal{T}}$ denotes the topology induced by $\bigoplus_{\alpha\in I} E_\alpha(\mathcal{T}_\alpha)$ on $\bigoplus_{\alpha\in I'} E_\alpha$.

(ii) The projection $\hat{P}_{I'} : \bigoplus_{\alpha\in I} E_\alpha(\mathcal{T}_\alpha) \longrightarrow \bigoplus_{\alpha\in I'} E_\alpha(\mathcal{T}_\alpha)$ is continuous, and hence $\bigoplus_{\alpha\in I'} E_\alpha(\mathcal{T}_\alpha)$ is a closed subspace of $\bigoplus_{\alpha\in I} E_\alpha(\mathcal{T}_\alpha)$.

Proof. (i) is a consequence of 4.(2), since for all U_n of the form (*) we have

$$U_n \cap (\bigoplus_{\alpha\in I'} E_\alpha) = (\sum_{k=1}^{\infty} \{ \bigcup_{\alpha\in I} U^\alpha_{2^{n-1}k} \}) \cap (\bigoplus_{\alpha\in I'} E_\alpha)$$
$$= \sum_{k=1}^{\infty} \{ \bigcup_{\alpha\in I'} U^\alpha_{2^{n-1}k} \} .$$

Hence the induced topology and the topology of $\bigoplus_{\alpha\in I'} E_\alpha(\mathcal{T}_\alpha)$ coincide. Furthermore we have

$$\hat{P}_{I'}(U_n) = \sum_{k=1}^{\infty} \{ \bigcup_{\alpha\in I'} U^\alpha_{2^{n-1}k} \} .$$

Hence $\hat{P}_{I'}$ is continuous.

A corollary of (1) is

(2) For all $\alpha \in I$ the space $E_\alpha(\mathfrak{T}_\alpha)$ is a closed subspace of $\bigoplus_{\alpha\in I} E_\alpha(\mathfrak{T}_\alpha)$.

Let $\mathfrak{T}_\pi$ denote again the topology induced by $\prod_{\alpha\in I} E_\alpha(\mathfrak{T}_\alpha)$ on its subspace $\bigoplus_{\alpha\in I} E_\alpha$.

(3) For a finite subset e of I the topologies $\mathfrak{T}$ and $\mathfrak{T}_\pi$ coincide on $\bigoplus_{\alpha\in e} E_\alpha$, if $E(\mathfrak{T}) = \bigoplus_{\alpha\in I} E_\alpha(\mathfrak{T}_\alpha)$.

Proof. Let k be the number of elements in e and let $\mathfrak{U} = (U_n)$ be a topological string in $\bigoplus_{\alpha\in I} E_\alpha(\mathfrak{T}_\alpha)$. $\mathfrak{U}_\alpha = (U_n^\alpha)$ with $U_n^\alpha := U_{k+n} \cap E_\alpha$ is a topological string in $E_\alpha(\mathfrak{T}_\alpha)$, $\alpha \in I$. Then $\mathfrak{W} = \bigcap_{\alpha\in e} P_\alpha^{-1}(\mathfrak{U}_\alpha)$ is by 3.(1) a topological string in $\prod_{\alpha\in I} E_\alpha(\mathfrak{T}_\alpha)$, where P_α denotes the projection of $\prod_{\alpha\in I} E_\alpha$ on E_α. For a knot $V_n \in \mathfrak{W}$ we have

$$V_n \cap (\bigoplus_{\alpha\in e} E_\alpha) \subset \bigoplus_{\alpha\in e} (U_{k+n} \cap E_\alpha) \subset (\sum_1^k U_{k+n}) \cap (\bigoplus_{\alpha\in e} E_\alpha)$$
$$\subset U_n \cap (\bigoplus_{\alpha\in e} E_\alpha).$$

Therefore $\mathfrak{T}|_{\bigoplus_{\alpha\in e} E_\alpha} \subset \mathfrak{T}_\pi|_{\bigoplus_{\alpha\in e} E_\alpha}$. Equality follows from $\mathfrak{T}_\pi \subset \mathfrak{T}$.

(4) If $E(\mathfrak{T}) = \bigoplus_{\alpha\in I} E_\alpha(\mathfrak{T}_\alpha)$, then $\mathfrak{T}$ and $\mathfrak{T}_\pi$ coincide, if and only if I is finite.

If I is finite, the assertion is contained in (3), and if I is infinite, one can easily see that $\mathfrak{T} \neq \mathfrak{T}_\pi$.

A further connection between the sum and the product topology gives the following proposition.

(5) If $E(\mathfrak{T}) = \bigoplus_{\alpha\in I} E_\alpha(\mathfrak{T}_\alpha)$, then there is in $E(\mathfrak{T})$ a base of neighbourhoods of 0 of $\mathfrak{T}_\pi$-closed sets.

Proof. Choose a knot U_n of the form (*) in $E(\mathfrak{T})$. We prove that $\overline{U_n}^{\mathfrak{T}}$ is $\mathfrak{T}_\pi$-closed. For this choose $x = (x_\alpha) \in \overline{\overline{U_n}^{\mathfrak{T}}}^{\mathfrak{T}_\pi}$ and let e be the finite set $e := \{\alpha : \alpha \in I \text{ with } x_\alpha \neq 0\}$. Then $x = P_e x \in \overline{P_e(\overline{U_n}^{\mathfrak{T}})}^{\mathfrak{T}_\pi} \subset \bigoplus_{\alpha\in e} E_\alpha(\widehat{\mathfrak{T}_\pi})$. Since $\bigoplus_{\alpha\in e} E_\alpha(\widehat{\mathfrak{T}_\pi}) = \bigoplus_{\alpha\in e} E_\alpha(\widehat{\mathfrak{T}})$, it follows $\overline{P_e(\overline{U_n}^{\mathfrak{T}})}^{\mathfrak{T}_\pi} = \overline{P_e(\overline{U_n}^{\mathfrak{T}})}^{\mathfrak{T}} \subset \overline{U_n}^{\mathfrak{T}}$ and $x \in \overline{U_n}^{\mathfrak{T}}$.

In the sequel we will consider a special class of inductive limits, which has some useful properties.

Let E be a vector space and let (E_k) be an ascending sequence of subspaces of E, such that $E = \bigcup_{k=1}^{\infty} E_k$. Furthermore we assume, that every E_k has a Hausdorff linear topology $\mathfrak{T}_k$, such that $\mathfrak{T}_{k+1}$ induces on E_k the topology $\mathfrak{T}_k$, i.e. $\mathfrak{T}_k = \widehat{\mathfrak{T}_{k+1}}$. Let $\mathfrak{T}$ denote the inductive topology on E with respect to the canonical embeddings $I_k : E_k(\mathfrak{T}_k) \longrightarrow E$, $k \in \mathbb{N}$.

By 4.(1) the set $\mathcal{F}$ of strings in E with $\mathcal{F} := \{\mathfrak{U} : \mathfrak{U} = (U_n)$ is a string in E, such that $(U_n \cap E_k)$ is for all $k \in \mathbb{N}$ a topological string in $E_k(\mathfrak{T}_k)\}$ forms a fundamental set of strings in $E(\mathfrak{T})$.

(6) $\mathfrak{T}$ <u>induces on every</u> E_k <u>the topology</u> $\mathfrak{T}_k$, <u>i.e.</u> $E_k(\mathfrak{T}_k) = E_k(\widehat{\mathfrak{T}})$.

Proof. If $\widehat{\mathfrak{T}}$ is the topology induced by $\mathfrak{T}$ on E_k, we have $\widehat{\mathfrak{T}} \subset \mathfrak{T}_k$. We have to show, that every $\mathfrak{T}_k$-string $\mathfrak{U} = (U_n)$ contains a $\widehat{\mathfrak{T}}$-string.

For $\mathfrak{U}$ there is a $\mathfrak{T}_k$-string $\mathfrak{W} = (W_n)$, such that for all $n \in \mathbb{N}$

$$\sum_{1}^{k+1} W_n \subset U_n .$$

For $1 \leq i \leq k$ the strings $\mathfrak{V}^i = (V_n^i)$ with $V_n^i := W_n \cap E_i$ are topological in $E_i(\mathfrak{T}_i)$. For $i > k$ we choose topological strings $\mathfrak{V}^i = (V_n^i)$ in $E_i(\mathfrak{T}_i)$, such that $(V_n^i + V_n^i) \cap E_{i-1} \subset V_n^{i-1}$ holds for all n .

The strings $\mathfrak{V}^i$, $i > k$, have the following property:

$$E_k \cap \Big(\sum_{i=k+1}^{k+j} V_n^i\Big) \subset E_k \cap (V_n^{k+1} + V_n^{k+1}) \subset V_n^k = W_n , \quad j \geq 1. \qquad (**)$$

We have to prove the first inclusion for $j \geq 2$:

$$E_k \cap \left(\sum_{i=k+1}^{k+j} V_n^i\right) \subset E_k \cap \left(\sum_{i=k+1}^{k+j} V_n^i + V_n^{k+j}\right)$$

$$= E_k \cap \left(E_{k+j-1} \cap \left\langle \left(\sum_{i=k+1}^{k+j-1} V_n^i\right) + V_n^{k+j} + V_n^{k+j} \right\rangle\right)$$

$$\subset E_k \cap \left(\sum_{i=k+1}^{k+j-1} V_n^i + \left\langle (V_n^{k+j} + V_n^{k+j}) \cap E_{k+j-1} \right\rangle\right)$$

$$\subset E_k \cap \left(\sum_{i=k+1}^{k+j-1} V_n^i + V_n^{k+j-1}\right) .$$

A continuation of these considerations shows that (**) is true.

The string $\mathfrak{W} = (V_n)$ in E with $V_n := \sum_{i=1}^{\infty} V_n^i$ is contained in $\mathfrak{F}$, hence $\mathfrak{W}$ is $\mathfrak{T}$-topological. For every knot $V_n \in \mathfrak{W}$ we have $V_n \cap E_k \subset U_n$:

$$E_k \cap V_n = E_k \cap \sum_{i=1}^{\infty} V_n^i = E_k \cap \left(\sum_{i=1}^{k} V_n^i + \sum_{i=k+1}^{\infty} V_n^i\right)$$

$$\subset \sum_{i=1}^{k} V_n^i + \left(E_k \cap \sum_{i=k+1}^{\infty} V_n^i\right)$$

$$\subset \sum_{i=1}^{k} V_n^i + \left(\bigcup_{j=1}^{\infty} \left\langle E_k \cap \sum_{i=k+1}^{k+j} V_n^i \right\rangle\right)$$

$$\subset \sum_{i=1}^{k} V_n^i + W_n \quad \text{by (**)} .$$

This gives

$$E_k \cap V_n \subset \sum_{1}^{k+1} W_n \subset U_n .$$

Hence $(V_n \cap E_k) \subset (U_n) = \mathfrak{U}$, and $(V_n \cap E_k)$ is a topological string in $E_k(\mathfrak{T})$.

As a corollary we get

(7) <u>Let</u> $E(\mathfrak{T})$ <u>be as in (6). Then</u> $\mathfrak{T}$ <u>is Hausdorff.</u>

If $E(\mathfrak{T})$ is as in (6), it is by (7) an inductive limit of the $E_k(\mathfrak{T}_k)$, that means $E(\mathfrak{T}) = \sum_{k=1}^{\infty} I_k(E_k(\mathfrak{T}_k))$. We call such an inductive limit $E(\mathfrak{T})$ a *strict inductive limit*, and we denote it by $E(\mathfrak{T}) = \bigcup_{k=1}^{\infty} E_k(\mathfrak{T}_k)$.

In the preceding propositions we have shown

(8) If $E(\mathfrak{T}) = \bigcup_{k=1}^{\infty} E_k(\mathfrak{T}_k)$, then the topology $\mathfrak{T}$ induces on each E_k the topology $\mathfrak{T}_k$.
If E_k is a closed subspace of $E_{k+1}(\mathfrak{T}_{k+1})$ for all $k \in \mathbb{N}$, then each E_k is also closed in $E(\mathfrak{T})$.

(9) Let $E(\mathfrak{T}) = \bigcup_{k=1}^{\infty} E_k(\mathfrak{T}_k)$ be the strict inductive limit of the $E_k(\mathfrak{T}_k)$ and let E_k be closed in $E_{k+1}(\mathfrak{T}_{k+1})$ for all $k \in \mathbb{N}$. A subset M of $E(\mathfrak{T})$ is bounded (precompact), if and only if M is contained in some E_k and bounded (precompact) with respect to $\mathfrak{T}_k$.

Proof. By (8) the condition is sufficient. Suppose M is bounded in $E(\mathfrak{T})$. If M is contained in no E_k, there are elements $x_k \in \frac{1}{k}M$ with $x_k \notin E_k$. Because E_k is closed in $E(\mathfrak{T})$ by (8), we find $\mathfrak{T}$-topological strings $\mathfrak{U}^k = (U_n^k)$, such that $x_k \notin E_k + U_1^k$. The string $\mathfrak{U} = (U_n)$ with $U_n := \bigcap_{k=1}^{\infty} (E_k + U_n^k)$ is topological in $E(\mathfrak{T})$. This follows from

$$\begin{aligned} E_j \cap U_n &= E_j \cap \bigcap_{k=1}^{\infty} (E_k + U_n^k) \\ &= (E_j \cap \bigcap_{k=1}^{j-1} (E_k + U_n^k)) \cap \bigcap_{k=j}^{\infty} (E_k + U_n^k) \\ &= E_j \cap \bigcap_{k=1}^{j-1} (E_k + U_n^k) \ . \end{aligned}$$

That means $\mathfrak{U}$ induces on each E_j a $\widehat{\mathfrak{T}}$- and therefore $\mathfrak{T}_j$-topological string. $\mathfrak{U}$ is topological, but U_1 contains no element of the 0-sequence (x_n).

Every countable topological direct sum $E(\mathfrak{T}) = \bigoplus_{i=1}^{\infty} E_i(\mathfrak{T}_i)$ is the strict inductive limit of the spaces $F_k(\mathfrak{T}'_k) := \bigoplus_{i=1}^{k} E_i(\mathfrak{T}_i)$, where F_k is closed in $F_{k+1}(\mathfrak{T}'_{k+1})$. We need this remark for the proof of the following proposition.

(10) <u>Assume</u> $E(\mathfrak{T}) = \bigoplus_{\alpha\in I} E_\alpha(\mathfrak{T}_\alpha)$. <u>A subset</u> M <u>of</u> $E(\mathfrak{T})$ <u>is bounded (precompact), if and only if there are a finite subset</u> $\{\alpha_1,\ldots,\alpha_m\}\subset I$ <u>and bounded (precompact) sets</u> $M_{\alpha_i}\subset E_{\alpha_i}(\mathfrak{T}_{\alpha_i})$, $1\le i\le n$, <u>such that</u> $M\subset\sum_{i=1}^{m} M_{\alpha_i}$.

<u>Proof.</u> Let M be bounded in $E(\mathfrak{T})$ and let $\hat{P}_\alpha$ be the projection from E onto E_α for all $\alpha\in I$. Then $\hat{P}_\alpha(M)\neq\{0\}$ for an at most finite set of $\alpha\in I$. Otherwise there would be a countable subset $I_0 = \{\alpha_1,\alpha_2,\ldots\}\subset I$, such that $\hat{P}_{I_0}(M)$ is bounded in $\bigoplus_{\alpha\in I_0} E_\alpha(\mathfrak{T}_\alpha) = \bigcup_{k=1}^{\infty}(\bigoplus_{j=1}^{k} E_{\alpha_j}(\mathfrak{T}_{\alpha_j}))$. Furthermore $\hat{P}_{I_0}(M)\not\subset\bigoplus_{j=1}^{k} E_{\alpha_j}(\mathfrak{T}_{\alpha_j})$ for all $k\in\mathbb{N}$. But this contradicts (9).

If $\hat{P}_{\alpha_i}(M)\neq\{0\}$ for $\alpha_1,\ldots,\alpha_m\in I$, we obtain $M\subset\sum_{i=1}^{m}\hat{P}_{\alpha_i}(M)$, where $\hat{P}_{\alpha_i}(M)$ is bounded by (1), (ii).

In general an inductive limit of complete spaces needs not be complete, but in the strict case we have (for a more general discussion of this question see the end of § 16):

(11) <u>A strict inductive limit</u> $E(\mathfrak{T}) = \bigcup_{k=1}^{\infty} E_k(\mathfrak{T}_k)$ <u>of complete spaces</u> $E_k(\mathfrak{T}_k)$ <u>is complete.</u>

<u>Proof.</u> Assume $E(\mathfrak{T})$ is not complete and let $\tilde{E}(\tilde{\mathfrak{T}})$ be the completion of $E(\mathfrak{T})$. If $x\in\tilde{E}\setminus E$, then $x\notin E_k$ for all $k\in\mathbb{N}$. E_k is closed in $\tilde{E}(\tilde{\mathfrak{T}})$ by (8). Hence there is a sequence $(\mathfrak{U}^k)$ of topological strings $\mathfrak{U}^k = (U_n^k)$ in $E(\mathfrak{T})$ with $(x+\tilde{U}_1^k)\cap E_k = \emptyset$, i.e. with $x\notin E_k+\tilde{U}_1^k$ *) for all k. Set $U_n := \bigcap_{k=1}^{\infty}(E_k+U_n^k)$. Then (U_n) is a topological string in $E(\mathfrak{T})$ (see the proof of (9)) and $(\tilde{U}_n)$ is topological in $\tilde{E}(\tilde{\mathfrak{T}})$. Since $E=\bigcup_{k=1}^{\infty}E_k$ is dense in $\tilde{E}(\tilde{\mathfrak{T}})$, we have

*) $\tilde{M}$ denotes the closure of M in $\tilde{E}(\tilde{\mathfrak{T}})$.

$(x+\tilde{U}_2) \cap E_{k_o} \neq \emptyset$ for a $k_o \in \mathbb{N}$ and $x \in E_{k_o} + \tilde{U}_2 \subset E_{k_o} + \widetilde{E_{k_o} + U_2^{k_o}} \subset E_{k_o} + \tilde{U}_1^{k_o}$. This is a contradiction.

(11) implies that for a countable direct sum $E(\mathfrak{f}) = \bigoplus_{k=1}^{\infty} E_k(\mathfrak{f}_k)$ the completeness of the $E(\mathfrak{f})$ follows from the completeness of the $E_k(\mathfrak{f}_k)$, but we have more general:

(12) <u>The topological direct sum</u> $E(\mathfrak{f}) = \bigoplus_{\alpha \in I} E_\alpha(\mathfrak{f}_\alpha)$ <u>of complete spaces</u> $E_\alpha(\mathfrak{f}_\alpha)$ <u>is complete.</u>

<u>Proof.</u> Let $\mathfrak{F}$ be a base of a Cauchy filter in $E(\mathfrak{f})$. Then $\hat{P}_\alpha(\mathfrak{F})$ is for $\alpha \in I$ a base of a Cauchy filter in $E_\alpha(\mathfrak{f}_\alpha)$ by (1), (ii) and $\hat{P}_\alpha(\mathfrak{F}) \longrightarrow x_\alpha \in E_\alpha$. We have $x_\alpha \neq 0$ for at most finitely many $\alpha \in I$ (this follows, since $\bigoplus_{n=1}^{\infty} E_{\alpha_n}(\mathfrak{f}_{\alpha_n})$ is complete for each subsequence $(\alpha_n) \subset I$) and therefore $x := (x_\alpha) \in E$. Furthermore $\mathfrak{F} \longrightarrow x$ in $E(\mathfrak{f}_\pi)$. But $\mathfrak{f} \supset \mathfrak{f}_\pi$ and $\mathfrak{f}$ has by (5) a base of neighbourhoods of 0 of $\mathfrak{f}_\pi$-closed sets, hence $\mathfrak{F} \longrightarrow x$ in $E(\mathfrak{f})$.

References: S.O. IYAHEN [1], J. KÖHN [1].

§ 6 Barrelled topological vector spaces

In every t.v.s. $E(\mathcal{T})$ there is a fundamental set of c l o s e d s t r i n g s (that are strings, whose knots are closed). But in general not every closed string in $E(\mathcal{T})$ is a topological string.

Example. We consider the space φ of all finite sequences as a subspace of the space ω of all sequences, ω endowed with the usual product topology. The set $U := \{(x_i) : (x_i) \in \varphi \text{ and } |x_i| \leq 1 \text{ for } i \in \mathbb{N}\}$ is absolutely convex, absorbing and closed in φ. The natural string $\mathcal{U} = (U_n)$ of U with $U_n := \frac{1}{2^{n-1}} U$ is a closed string in φ, but the U_n are not 0-neighbourhoods in the topology induced on φ by ω.

A t.v.s. $E(\mathcal{T})$ is called b a r r e l l e d i n $\mathcal{L}$ (or $\mathcal{L}$-b a r r e l l e d) *), if all closed strings in $E(\mathcal{T})$ are topological strings.

Any vector space endowed with its finest linear topology $\mathcal{T}^f$ is barrelled. More examples are provided by the following proposition:

(1) Every t.v.s. $E(\mathcal{T})$ of second category (i.e. a Baire space) is barrelled. Especially (F)-spaces are barrelled.

Proof. Let $\mathcal{U} = (U_n)$ be a closed string in $E(\mathcal{T})$. For fixed $n \in \mathbb{N}$ we have $E = \bigcup_{k=1}^{\infty} kU_{n+1}$. $E(\mathcal{T})$ is of second category and so U_{n+1} has an inner point. Since $U_{n+1} + U_{n+1} \subset U_n$, the knot U_n has 0 as in-

---

*) We will omit "in $\mathcal{L}$", if we consider only the category $\mathcal{L}$ of t.v.s. without looking at relations to the category of locally convex spaces.

ner point and is a neighbourhood of 0. Hence $\mathfrak{U}$ is topological.

A consequence of (1) is, that all examples in § 2 are barrelled spaces.

Before considering permanence properties of barrelled spaces we will prove two characteristic properties of these spaces.

By 1.(2) the closed strings in a t.v.s. $E(\mathcal{T})$ generate a linear topology on E. We call this topology the s t r o n g t o p o l o g y of $E(\mathcal{T})$ and denote it by $\mathcal{T}^b$. We have $\mathcal{T} \subset \mathcal{T}^b$. From this definition follows immediately

(2) <u>For a t.v.s.</u> $E(\mathcal{T})$ <u>the following conditions are equivalent:</u>

(i) $E(\mathcal{T})$ <u>is barrelled.</u>

(ii) <u>It is</u> $\mathcal{T} = \mathcal{T}^b$.

(iii) <u>Every Hausdorff linear topology</u> $\mathcal{T}'$ <u>on</u> E <u>which has a base of 0-neighbourhoods of</u> $\mathcal{T}$<u>-closed sets, is coarser than</u> $\mathcal{T}$.

Later on we will need

(3) <u>If</u> $A: E(\mathcal{T}_1) \longrightarrow F(\mathcal{T}_2)$ <u>is a continuous linear mapping, then</u> $A: E(\mathcal{T}_1{}^b) \longrightarrow F(\mathcal{T}_2{}^b)$ <u>is also continuous.</u>

We want to show some permanence properties of barrelled spaces.

(4) <u>If</u> $E(\mathcal{T}) = \sum_{\alpha \in I} A_\alpha(E_\alpha(\mathcal{T}_\alpha))$ <u>is the inductive limit of the barrelled spaces</u> $E_\alpha(\mathcal{T}_\alpha)$, <u>then</u> $E(\mathcal{T})$ <u>is also barrelled.</u>

<u>Proof.</u> Let $\mathfrak{U}$ be a closed string in $E(\mathcal{T})$. Then $A^{-1}(\mathfrak{U})$ is a closed, hence topological string in $E_\alpha(\mathcal{T}_\alpha)$ for all $\alpha \in I$. By 4.(1) $\mathfrak{U}$ is topological in $E(\mathcal{T})$.

A consequence of (4) is

(5) <u>The topological direct sum of barrelled spaces is barrelled. Every quotient space of a barrelled space is barrelled.</u>

A result similar to (4) for projective limits does not hold as the example at the beginning of this section shows (see also 3.(4)). However we can prove, that the product of barrelled spaces is barrelled. For this we need some preparations.

(6) <u>If</u> $\mathfrak{U} = (U_n)$ <u>is a closed string in the t.v.s.</u> $E(\mathcal{T})$, <u>then each</u> $U_n \in \mathfrak{U}$ <u>absorbs the Banach disks</u> K <u>in</u> $E(\mathcal{T})$. *)

<u>Proof.</u> We endow the vector space $E_K := \bigcup_{n=1}^{\infty} nK$ with the topology $\mathcal{T}_K$ generated by the norm $q_K(x) := \inf\{|\lambda| : x \in \lambda K\}$. $E_K(\mathcal{T}_K)$ is a Banach space and hence barrelled. The topology induced by $\mathcal{T}$ on E_K is coarser than $\mathcal{T}_K$. Hence the string $(U_n \cap E_K)$ is closed and therefore topological in $E_K(\mathcal{T}_K)$. That means every U_n absorbs the set K.

(7) <u>If</u> U <u>is a balanced set in the topological product</u> $\prod_{\alpha \in I} E_\alpha(\mathcal{T}_\alpha)$, <u>which absorbs every compact absolutely convex subset, then there is a finite subset</u> e <u>of</u> I, <u>such that</u> $U \supset \bigoplus_{\alpha \in I \setminus e} E_\alpha$. **)

<u>Proof.</u> It suffices to prove (7) for an infinite index set I. Assume the assertion is wrong. Then there is an element $x^1 = (x_\alpha^1) \in \bigoplus_{\alpha \in I} E_\alpha$, such that $x^1 \notin U$. Set $e_1 := \{\alpha \in I : x_\alpha^1 \neq 0\}$. Now there is an

*) A bounded absolutely convex subset K of $E(\mathcal{T})$ is called B a n a c h d i s k , if E_K is a Banach space with the norm generated by K. Special Banach disks are compact absolutely convex sets.

**) Here we consider again $\bigoplus_{\alpha \in I \setminus e} E_\alpha$ as a subspace of $\bigoplus_{\alpha \in I} E_\alpha$ or $\prod_{\alpha \in I} E_\alpha$.

element $x^2 = (x^2_\alpha) \in \bigoplus_{\alpha\in I\setminus e_1} E_\alpha$, such that $x^2 \notin 2U$. Set $e_2 := e_1 \cup \{\alpha \in I: x^2_\alpha \neq 0\}$. Continuing this construction we obtain a sequence (x^n) and a sequence (e_n), such that $x^n \notin nU$ and $x^n \in \bigoplus_{\alpha\in I\setminus e_{n-1}} E_\alpha$, where $e_{n-1} := e_{n-2} \cup \{\alpha \in I: x^{n-1}_\alpha \neq 0\}$ for $n \geq 3$.

The sets $[x^n]_1 = \{\lambda x^n: |\lambda| \leq 1\}$ are absolutely convex and compact in $\prod_{\alpha\in I} E_\alpha(\mathcal{T}_\alpha)$. Hence $\prod_{n=1}^{\infty} [x^n]_1$ is absolutely convex and compact. Since U absorbs compact absolutely convex sets, there is a $k \in \mathbb{N}$, such that $\prod_{n=1}^{\infty} [x^n]_1 \subset kU$. Hence $x^k \in kU$, which contradicts the choice of x^k.

Now we can prove

(8) <u>If</u> $E(\mathcal{T}) = \prod_{\alpha\in I} E_\alpha(\mathcal{T}_\alpha)$, <u>then</u> $E(\mathcal{T}^b) = \prod_{\alpha\in I} E_\alpha(\mathcal{T}^b_\alpha)$.

<u>Proof.</u> For finite sets I (8) follows from 5.(4). Hence assume I infinite. Let $\mathcal{T}'$ denote the product topology of $\prod_{\alpha\in I} E_\alpha(\mathcal{T}_\alpha^b)$. By (3) it is obvious that $\mathcal{T}' \subset \mathcal{T}^b$. It suffices to show that each closed string $\mathfrak{U} = (U_n)$ in $E(\mathcal{T})$ is topological in $E(\mathcal{T}') = \prod_{\alpha\in I} E_\alpha(\mathcal{T}_\alpha^b)$.

By (6) and (7) there is for every $k \in \mathbb{N}$ a finite subset $e_k \subset I$, such that $U_{k+1} \supset \bigoplus_{\alpha\in I\setminus e_k} E_\alpha$. Since U_{k+1} is closed, we have also $U_{k+1} \supset \prod_{\alpha\in I\setminus e_k} E_\alpha$. $(U_n \cap \prod_{\alpha\in e_k} E_\alpha)$ is a closed string in $\prod_{\alpha\in e_k} E_\alpha(\mathcal{T}_\alpha)$ and therefore $U_{k+1} \cap \prod_{\alpha\in e_k} E_\alpha$ is a 0-neighbourhood in $\prod_{\alpha\in e_k} E_\alpha(\mathcal{T}_\alpha^b)$. From this we obtain that $(U_{k+1} \cap \prod_{\alpha\in e_k} E_\alpha) + \prod_{\alpha\in I\setminus e_k} E_\alpha$ is a 0-neighbourhood in the space $\prod_{\alpha\in I} E_\alpha(\mathcal{T}_\alpha^b)$. Since $U_k \supset U_{k+1} + U_{k+1} \supset (U_{k+1} \cap \prod_{\alpha\in e_k} E_\alpha) + \prod_{\alpha\in I\setminus e_k} E_\alpha$, U_k is a 0-neighbourhood in $\prod_{\alpha\in I} E_\alpha(\mathcal{T}_\alpha^b)$.

As a corollary we obtain by (2)

(9) <u>The product</u> $\prod_{\alpha\in I} E_\alpha(\mathcal{T}_\alpha)$ <u>of barrelled spaces</u> $E_\alpha(\mathcal{T}_\alpha)$ <u>is barrelled.</u>

Let H be a linear subspace of the t.v.s. $E(\mathfrak{T})$ and let $\hat{\mathfrak{T}}$ denote the topology induced on H by $\mathfrak{T}$.

(10) *If* H *is a subspace of finite codimension in* $E(\mathfrak{T})$, *then we have* $\hat{\mathfrak{T}}^b = \widehat{\mathfrak{T}^b}$.

Proof. It suffices to consider the case of a hyperplane in E. $\widehat{\mathfrak{T}^b} \leq \hat{\mathfrak{T}}^b$ is obvious. We only have to show that every closed string $\mathfrak{U} = (U_n)$ in $H(\hat{\mathfrak{T}})$ is topological in $H(\widehat{\mathfrak{T}^b})$. There are two possible cases:

(i) No knot of $\mathfrak{U} = (U_n)$ is closed in $E(\mathfrak{T})$. Then the closures $\overline{U_n}$ in $E(\mathfrak{T})$ are the knots of a closed string $(\overline{U_n})$ in $E(\mathfrak{T})$. Hence $(\overline{U_n})$ is topological in $E(\mathfrak{T}^b)$, and (U_n) is topological in $H(\widehat{\mathfrak{T}^b})$, since $U_n = \overline{U_n} \cap H$.

(ii) There is a knot U_n in $\mathfrak{U}$, which is closed in $E(\mathfrak{T})$. Then all the knots U_{n+i} for $i \in \mathbb{N}$ are also closed in $E(\mathfrak{T})$. Choose $x \in E$ with $x \notin H$. The set $[x]_1 = \{\lambda x: |\lambda| \leq 1\}$ is compact in $E(\mathfrak{T})$ and therefore $U_{n+i} + \frac{1}{2^i}[x]_1$ is closed in $E(\mathfrak{T})$ for all $i \in \mathbb{N}$. Since now $U_{n+i} + \frac{1}{2^i}[x]_1$ is also absorbing in E, $(U_{n+i} + \frac{1}{2^i}[x]_1)_{i \in \mathbb{N}}$ is a topological string in $E(\mathfrak{T}^b)$. With $(U_{n+i} + \frac{1}{2^i}[x]_1) \cap H = U_{n+i}$ we obtain that $\mathfrak{U}$ is topological in $H(\widehat{\mathfrak{T}^b})$.

From (10) follows

(11) *Let* H *be a linear subspace of finite codimension in the barrelled space* $E(\mathfrak{T})$. *Then* $H(\hat{\mathfrak{T}})$ *is also barrelled.*

(11) is also true for subspaces of countable codimension (see the end of § 16).

We omit the easy proof of the following proposition:

(12) If $E(\mathfrak{T})$ contains a dense linear subspace H, such that $H(\mathfrak{T})$ is barrelled, then $E(\mathfrak{T})$ is barrelled.
Especially the completion of a barrelled space is barrelled.

If $E(\mathfrak{T})$ is a t.v.s., then the finest linear topology $\mathfrak{T}^f$ on E is barrelled with $\mathfrak{T}^f \supset \mathfrak{T}$. Among all barrelled topologies on E which are finer than $\mathfrak{T}$, there is a coarsest topology $\mathfrak{T}^t$ by (4). We call $\mathfrak{T}^t$ the a s s o c i a t e d b a r r e l l e d t o p o l o g y of $\mathfrak{T}$. We have $\mathfrak{T} \subset \mathfrak{T}^b \subset \mathfrak{T}^t$ ($\mathfrak{T}^t$ can also be obtained from $\mathfrak{T}$ by a "transfinite iteration" of strong topologies).

As in (3) we have

(13) If the linear mapping $A: E(\mathfrak{T}_1) \longrightarrow F(\mathfrak{T}_2)$ is continuous, then $A: E(\mathfrak{T}_1^t) \longrightarrow F(\mathfrak{T}_2^t)$ is continuous too.

Proof. Among all linear topologies $\mathfrak{T}'$ on F, for which the mapping $A: E(\mathfrak{T}_1^t) \longrightarrow F(\mathfrak{T}')$ is continuous, there is a finest one. Denote it by $\mathfrak{T}$. Then $\mathfrak{T} = \mathfrak{T}^b$ by (3), because $(\mathfrak{T}_1^t)^b = \mathfrak{T}_1^t$. Hence $\mathfrak{T}$ is barrelled. $\mathfrak{T} \supset \mathfrak{T}_2$ implies $\mathfrak{T} \supset \mathfrak{T}_2^t$, and $A: E(\mathfrak{T}_1^t) \longrightarrow F(\mathfrak{T}_2^t)$ is continuous.

If the t.v.s. $E(\mathfrak{T})$ is complete, then $E(\mathfrak{T}^b)$ is complete (see G. KÖTHE [4], §18,4.(4)), but $E(\mathfrak{T}^t)$ is also complete as we want to show.

We recall the f i l t e r c o n d i t i o n of W. ROBERTSON [1]. Assume there is given a finer linear topology $\mathfrak{T}_1$ on the t.v.s. $E(\mathfrak{T})$. The identity $I: E(\mathfrak{T}_1) \longrightarrow E(\mathfrak{T})$ satisfies the filter condition, if every Cauchy filter in $E(\mathfrak{T}_1)$, which converges in $E(\mathfrak{T})$ to an x, converges in $E(\mathfrak{T}_1)$ to x. The filter condition holds for $I: E(\mathfrak{T}_1) \longrightarrow E(\mathfrak{T})$, if and only if the continuous extension $\tilde{I}: \tilde{E}(\tilde{\mathfrak{T}}_1) \longrightarrow \tilde{E}(\tilde{\mathfrak{T}})$ of I onto the completion $\tilde{E}(\tilde{\mathfrak{T}}_1)$ is one-to-one.

If the filter condition holds for $I: E(\mathfrak{T}_1) \longrightarrow E(\mathfrak{T})$, then it holds for $I: E(\mathfrak{T}_1^b) \longrightarrow E(\mathfrak{T})$ (G. KÖTHE [4], §18,4.(4)).

If we set $\mathfrak{T}_1$ for the supremum of all linear topologies $\mathfrak{T}'$ on E with $\mathfrak{T} \subset \mathfrak{T}' \subset \mathfrak{T}^t$, such that the filter condition holds for $I: E(\mathfrak{T}') \longrightarrow E(\mathfrak{T})$, then the filter condition holds for $I: E(\mathfrak{T}_1) \longrightarrow$

$E(\mathfrak{T})$, and therefore for $I: E(\mathfrak{T}_1^b) \longrightarrow E(\mathfrak{T})$ too. Hence $\mathfrak{T}_1 = \mathfrak{T}_1^b$ and $\mathfrak{T}_1$ is barrelled. Therefore $\mathfrak{T}_1 = \mathfrak{T}^t$ and we have

(14) <u>If</u> $E(\mathfrak{T})$ <u>is a t.v.s., then:</u>

(i) <u>The filter condition holds for the identity mapping</u> $I: E(\mathfrak{T}^t) \longrightarrow E(\mathfrak{T})$.

(ii) <u>The continuous extension</u> $\tilde{I}: \tilde{E}(\widetilde{\mathfrak{T}^t}) \longrightarrow \tilde{E}(\tilde{\mathfrak{T}})$ <u>of</u> I <u>is a one-to-one mapping.</u>

(iii) <u>If</u> $E(\mathfrak{T})$ <u>is complete, then</u> $E(\mathfrak{T}^t)$ <u>is complete.</u>

For the results of this section see N. ADASCH [2],[5], S.O. IYAHEN [1], W. ROBERTSON [1], S. TOMASEK [1].

§ 7 The Banach-Steinhaus theorem

We want to give another characterization of barrelled spaces in this section, a generalization of the classical theorem of Banach-Steinhaus.

Let $E(\mathfrak{T}_1)$ and $F(\mathfrak{T}_2)$ be two t.v.s. and let $\mathfrak{A}$ be an equicontinuous set of linear mappings from $E(\mathfrak{T}_1)$ into $F(\mathfrak{T}_2)$. Then $\mathfrak{A}$ is pointwise bounded, i.e. for all $x \in E$ the set $\{Ax: A \in \mathfrak{A}\}$ is bounded in $F(\mathfrak{T}_2)$.

Now consider the class of t.v.s. $E(\mathfrak{T}_1)$ with the property that pointwise bounded sets of continuous linear mappings are equicontinuous. We will show that this class is exactly the class of barrelled spaces.

(1) <u>Let $E(\mathfrak{T}_1)$ be a t.v.s. and let $\mathfrak{A}$ be a pointwise bounded set of continuous linear mappings from $E(\mathfrak{T}_1)$ into another t.v.s. $F(\mathfrak{T}_2)$. Then $\mathfrak{A}$ is equicontinuous with respect to $E(\mathfrak{T}_1^b)$ and $F(\mathfrak{T}_2)$.</u>

<u>Proof.</u> If $\mathfrak{W}$ is a closed topological string in $F(\mathfrak{T}_2)$, then $\mathfrak{U} := \bigcap_{A \in \mathfrak{A}} A^{-1}(\mathfrak{W})$ is a closed string in $E(\mathfrak{T}_1)$, hence a topological string in $E(\mathfrak{T}_1^b)$. On the other hand $A(\mathfrak{U}) \subset \mathfrak{W}$ for all $A \in \mathfrak{A}$.

(2) <u>Let $E(\mathfrak{T}_1)$ be a t.v.s. such that every pointwise bounded set $\mathfrak{A}$ of continuous linear mappings from $E(\mathfrak{T}_1)$ into any (F)-space $F(\mathfrak{T}_2)$ is equicontinuous. Then $\mathfrak{T}_1 = \mathfrak{T}_1^b$ holds, i.e. $\mathfrak{T}_1$ is barrelled.</u>

<u>Proof.</u> We have to show that every closed string $\mathfrak{U}$ in $E(\mathfrak{T}_1)$ is topological. Let $\{\mathfrak{W}^\gamma\}_{\gamma \in \Gamma}$ be a set of topological strings $\mathfrak{W}^\gamma = (V_n^\gamma)$ in $E(\mathfrak{T}_1)$, such that $\{V_1^\gamma: \gamma \in \Gamma\}$ is a base of neighbourhoods of 0 .

If $E_\gamma := E$ for all $\gamma \in \Gamma$, we form the algebraic direct sum $F := \bigoplus_{\gamma\in\Gamma} E_\gamma$. By I_γ we denote the embeddings of E_γ into F.

Let $\mathfrak{U} = (U_n)$ be a closed string in $E(\mathcal{T}_1)$. Then the sets $W_n := \bigoplus_{\gamma\in\Gamma}(U_n + V_n^\gamma)$, $n \in \mathbb{N}$, form a string $\mathfrak{M} = (W_n)$ in F. Hence $F_{\mathfrak{M}}$ is a metrizable t.v.s.. Let K be the canonical mapping from F into $\widetilde{F_{\mathfrak{M}}}$.

Since $I_\gamma(V_n^\gamma) \subset \bigoplus_{\gamma\in\Gamma}(U_n + V_n^\gamma)$, the mappings $K \circ I_\gamma$ are continuous mappings from $E(\mathcal{T}_1)$ into $\widetilde{F_{\mathfrak{M}}}$. The set $\{K \circ I_\gamma\}_{\gamma\in\Gamma}$ is also pointwise bounded: for $x \in E$ and $x \in \lambda U_n$ we have $\{I_\gamma(x)\}_{\gamma\in\Gamma} \subset \lambda(\bigoplus_{\gamma\in\Gamma}(U_n + V_n^\gamma))$.

By assumption $\{K \circ I_\gamma\}_{\gamma\in\Gamma}$ must be an equicontinuous set of mappings from $E(\mathcal{T}_1)$ into $\widetilde{F_{\mathfrak{M}}}$. Therefore the sets

$$\bigcap_{\gamma'\in\Gamma}(K \circ I_{\gamma'})^{-1}\left[\overline{K(\bigoplus_{\gamma\in\Gamma}(U_n + V_n^\gamma))}\right]$$

are 0-neighbourhoods in $E(\mathcal{T}_1)$ for all $n \in \mathbb{N}$. But on the other hand

$$\begin{aligned}\bigcap_{\gamma'\in\Gamma}(K \circ I_{\gamma'})^{-1}\left[\overline{K(\bigoplus_{\gamma\in\Gamma}(U_{n+2} + V_{n+2}^\gamma))}\right] &\subset \bigcap_{\gamma'\in\Gamma} I_{\gamma'}^{-1}(\bigoplus_{\gamma\in\Gamma}(U_n + V_n^\gamma)) \\ &= \bigcap_{\gamma'\in\Gamma}(U_n + V_n^{\gamma'}) \\ &= U_n\end{aligned}$$

holds, since U_n is closed. Hence $\mathfrak{U}$ is topological in $E(\mathcal{T}_1)$.

Combining (1) and (2) we get

(3) <u>For a t.v.s.</u> $E(\mathcal{T})$ <u>the following conditions are equivalent:</u>

(i) $E(\mathcal{T})$ <u>is barrelled.</u>

(ii) <u>Every pointwise bounded set of continuous linear mappings from</u> $E(\mathcal{T})$ <u>into an arbitrary t.v.s. is equicontinuous.</u>

(iii) <u>Every pointwise bounded set of continuous linear mappings from</u> $E(\mathcal{T})$ <u>into an arbitrary (F)-space is equicontinuous.</u>

Since a metrizable t.v.s. has a countable base of neighbourhoods of 0, one gets with the same proof as (2)

(4) A metrizable t.v.s. $E(\mathcal{T})$ is barrelled, if and only if every pointwise bounded sequence of continuous linear mappings from $E(\mathcal{T})$ into an arbitrary (F)-space is equicontinuous.

An important consequence of the Banach-Steinhaus theorem (3) (often this consequence is called "Banach-Steinhaus theorem") says, that the pointwise limit of a sequence of continuous linear mappings from an (F)-space into an arbitrary t.v.s. is a continuous linear mapping. This assertion holds also for barrelled spaces.

(5) Let $E(\mathcal{T}_1)$ be a barrelled and $F(\mathcal{T}_2)$ an arbitrary t.v.s.. Let $\{A_\iota\}_{\iota\in I}$ be a directed system of continuous linear mappings from $E(\mathcal{T}_1)$ into $F(\mathcal{T}_2)$, which converges pointwise. If there is an $\iota_0 \in I$, such that $\{A_\iota(x): \iota \geq \iota_0\}$ is bounded in $F(\mathcal{T}_2)$ for every $x \in E$, then the pointwise limit $A: x \longrightarrow \lim A_\iota(x)$ is a continuous linear mapping from $E(\mathcal{T}_1)$ into $F(\mathcal{T}_2)$.

Proof. The set $\mathfrak{A} := \{A_\iota: \iota \geq \iota_0\}$ is pointwise bounded, hence $\mathfrak{A}$ is equicontinuous by (3). For every closed 0-neighbourhood V in $F(\mathcal{T}_2)$ there is a 0-neighbourhood U in $E(\mathcal{T}_1)$, such that $\bigcup_{\iota\geq\iota_0} A_\iota(U) \subset V$. Then the inclusions $A(U) \subset \overline{\bigcup_{\iota\geq\iota_0} A_\iota(U)} \subset V$ show the continuity of A.

(5) gives

(6) Let $E(\mathcal{T}_1)$ and $F(\mathcal{T}_2)$ be as in (5). Every pointwise limit of a sequence $(A_n)_{n\in\mathbb{N}}$ of continuous linear mappings from $E(\mathcal{T}_1)$ into $F(\mathcal{T}_2)$ is continuous.

References: D. KEIM [1], W. ROBERTSON [1], L. WAELBROECK [1].

§ 8 Barrelled spaces and the closed graph theorem

It is easy to see:

(1) *If A is a continuous mapping from the Hausdorff topological space $E(\mathfrak{T}_1)$ into the Hausdorff topological space $F(\mathfrak{T}_2)$, then the graph $G(A) := \{(x,Ax): x \in E\}$ of the mapping A is closed in the product $E(\mathfrak{T}_1) \times F(\mathfrak{T}_2)$.*

Assume $(x,y) \notin G(A)$, which means $A(x) \neq y$. Then there are neighbourhoods $V(y)$ and $V(Ax)$ such that $V(y) \cap V(Ax) = \emptyset$. $A^{-1}(V(Ax))$ is a neighbourhood of x by continuity of A. $A^{-1}(V(Ax)) \times V(y)$ is a neighbourhood of (x,y) which contains no point of G(A).

A mapping with a closed graph is called a c l o s e d m a p p i n g . (1) says therefore that continuous mappings are always closed mappings. In the class of topological vector spaces we now want to give conditions, such that conversely closed linear mappings are continuous. The theory which will be developped in this section is a direct generalization of BANACH's theory (cf. [1]). S. BANACH proved that every closed linear mapping from an (F)-space into an (F)-space is continuous. First we generalize this to

(2) *Every closed linear mapping from a barrelled space $E(\mathfrak{T}_1)$ into an (F)-space $F(\mathfrak{T}_2)$ is continuous.*

The following lemma is needed for the proof of (2).

(3) (i) A linear mapping A of the t.v.s. $E(\mathfrak{T}_1)$ into the t.v.s. $F(\mathfrak{T}_2)$ is closed, if and only if there exists a Hausdorff linear topology $\mathfrak{T}_3$ on F with $\mathfrak{T}_3 \subset \mathfrak{T}_2$, such that $A: E(\mathfrak{T}_1) \longrightarrow F(\mathfrak{T}_3)$ is continuous.

(ii) A is closed, if and only if the strings $A(\mathfrak{U}) + \mathfrak{W}$, $\mathfrak{U}$ a topological string in $E(\mathfrak{T}_1)$, $\mathfrak{W}$ a topological string in $F(\mathfrak{T}_2)$, generate a Hausdorff linear topology on F, i.e. if

$$\bigcap_{U,V} (A(U) + V) = \{0\}$$

(U resp. V 0-neighbourhoods of $E(\mathfrak{T}_1)$ resp. $F(\mathfrak{T}_2)$).

Proof. Let $\mathfrak{T}_3$ be a Hausdorff linear topology on F, $\mathfrak{T}_3$ coarser than $\mathfrak{T}_2$, such that $A: E(\mathfrak{T}_1) \longrightarrow F(\mathfrak{T}_3)$ is continuous. Then by (1) G(A) is closed in $E(\mathfrak{T}_1) \times F(\mathfrak{T}_3)$. Since the product topology of $E(\mathfrak{T}_1) \times F(\mathfrak{T}_2)$ is finer, G(A) is also closed in it.

Let on the other hand A be closed. Then $\bigcap_{U,V} (A(U) + V) = \{0\}$ (For every $y \in F$, $y \neq 0$, $(0,y)$ is not contained in G(A). Hence there is a neighbourhood (U,V) of 0 in $E(\mathfrak{T}_1) \times F(\mathfrak{T}_2)$, such that $((0,y) + (U,V)) \cap G(A) = \emptyset$. From this follows $y \notin A(U) + V$.). Hence the topology $\mathfrak{T}_3$ generated by the strings $A(\mathfrak{U}) + \mathfrak{W}$ is Hausdorff. $\mathfrak{T}_3$ is coarser than $\mathfrak{T}_2$ and $A: E(\mathfrak{T}_1) \longrightarrow F(\mathfrak{T}_3)$ is continuous.

Fundamental for (F)-spaces is the proposition

(4) On an (F)-space $F(\mathfrak{T}_0)$ there exists no coarser Hausdorff barrelled topology.

Proof. Assume $\mathfrak{T}$ is barrelled with $\mathfrak{T} \subset \mathfrak{T}_0$. Let $\mathfrak{U} = (U_n)$ be a closed string in $F(\mathfrak{T}_0)$ which generates $\mathfrak{T}_0$. We prove that $\mathfrak{U}$ is $\mathfrak{T}$-topological too, showing $\overline{U_{n+1}}^{\mathfrak{T}} \subset U_n$ for all $n \in \mathbb{N}$ ($\mathfrak{T}$ is barrelled, hence $(\overline{U_n}^{\mathfrak{T}})$ is $\mathfrak{T}$-topological).

Choose $x \in \overline{U_{n+1}}^{\mathfrak{T}}$, hence $(x + \overline{U_{n+2}}^{\mathfrak{T}}) \cap U_{n+1} \neq \emptyset$. There exists $x_1 \in U_{n+1}$ with $x - x_1 \in \overline{U_{n+2}}^{\mathfrak{T}}$. There exists $x_2 \in U_{n+2}$, such that we have

$x - x_1 - x_2 \in \overline{U_{n+3}}^{\mathfrak{T}}$, because $(x - x_1 + \overline{U_{n+3}}^{\mathfrak{T}}) \cap U_{n+2} \neq \emptyset$. In this way we get a sequence (x_j) with $x_j \in U_{n+j}$ and $x - \sum_{i=1}^{j} x_i \in \overline{U_{n+1+j}}^{\mathfrak{T}}$. The Cauchy sequence $(\sum_{i=1}^{j} x_i)$ converges to a y in $F(\mathfrak{T}_0)$. Since $\sum_{i=1}^{j} x_i \in \sum_{i=1}^{j} U_{n+i} \subset U_n$ and U_n is $\mathfrak{T}_0$-closed, we have $y \in U_n$.

Since $\mathfrak{T} \subset \mathfrak{T}_0$, $(\sum_{i=1}^{j} x_i)$ converges to y in $F(\mathfrak{T})$. From $x - \sum_{i=1}^{j} x_i \in \overline{U_{n+j+1}}^{\mathfrak{T}}$ for all j follows $x - y \in \bigcap_{j=1}^{\infty} \overline{U_j}^{\mathfrak{T}}$. But $\bigcap_{j=1}^{\infty} \overline{U_j}^{\mathfrak{T}} = \{0\}$ (if a $z \neq 0$, there is a 0-neighbourhood V in $F(\mathfrak{T})$ with $(z + V) \cap V = \emptyset$, hence $(z + V) \cap U_j = \emptyset$ for some j and $z \notin \overline{U_j}^{\mathfrak{T}}$), therefore $x = y \in U_n$.

Now we can give the proof of (2): Let $A: E(\mathfrak{T}_1) \longrightarrow F(\mathfrak{T}_2)$ be closed. By (3) there exists a Hausdorff linear topology $\mathfrak{T}_3 \subset \mathfrak{T}_2$, such that $A: E(\mathfrak{T}_1) \longrightarrow F(\mathfrak{T}_3)$ is continuous. Then by 6.(13) $A: E(\mathfrak{T}_1^t) \longrightarrow F(\mathfrak{T}_3^t)$ is also continuous. Since $\mathfrak{T}_3^t \subset \mathfrak{T}_2^t = \mathfrak{T}_2$ (cf. 6.(1)), we have $\mathfrak{T}_3^t = \mathfrak{T}_2$ by (4). Since $E(\mathfrak{T}_1)$ is barrelled, we have $\mathfrak{T}_1^t = \mathfrak{T}_1$ and hence $A: E(\mathfrak{T}_1) \longrightarrow F(\mathfrak{T}_2)$ is continuous.

We now prove the converse of (2) and give thus a new characterization of barrelled spaces (see M. MAHOWALD [1], S.O. IYAHEN [1]).

(5) Let $E(\mathfrak{T})$ be a t.v.s. such that every closed linear mapping from $E(\mathfrak{T})$ into an (F)-space is continuous. Then $E(\mathfrak{T})$ is barrelled.

Proof. Let $\mathfrak{U} = (U_n)$ be a closed string in $E(\mathfrak{T})$. Consider the (F)-space $\widetilde{E_{\mathfrak{U}}}$ (see §2) and the canonical mapping $K: E(\mathfrak{T}) \longrightarrow \widetilde{E_{\mathfrak{U}}}$. If K is closed, then K is continuous by assumption on $E(\mathfrak{T})$, and $\mathfrak{U}$ is a topological string in $E(\mathfrak{T})$, because $K^{-1}(\widetilde{K(U_{n+2})}) = K^{-1}(\widetilde{K(U_{n+2})} \cap E/N(\mathfrak{U})) \subset K^{-1}(K(U_{n+2}) + K(U_{n+2})) \subset U_{n+1} + N(\mathfrak{U}) \subset U_n$ ($\widetilde{K(U_n)}$ is the closure of $K(U_n)$ in $\widetilde{E_{\mathfrak{U}}}$).

We prove $G(K) = \overline{G(K)}$, showing that $(x,y) \in \overline{G(K)}$ implies $K(x) - y \in \bigcap_{n=1}^{\infty} \widetilde{K(U_n)} = \{0\}$. If $(x,y) \in \overline{G(K)}$, we have $[(x,y) + (U, \widetilde{K(U_{n+4})})] \cap$

$G(K) \neq \emptyset$ for all O-neighbourhoods U in $E(\mathfrak{T})$, hence $K(x) - y \in K(U) + \widetilde{K(U_{n+4})}$ for all U. There exists $z \in E$ with $K(z) \in y + \widetilde{K(U_{n+4})}$. For this z we have

$$K(x) - K(z) \in K(U) + \widetilde{K(U_{n+4})} + \widetilde{K(U_{n+4})} \subset K(U) + \widetilde{K(U_{n+3})} ,$$

$$K(x) - K(z) \in K(U) + (\widetilde{K(U_{n+3})} \cap E/N(\mathfrak{U})) \subset K(U) + K(U_{n+3}) + K(U_{n+3}),$$

$$x - z \in U + U_{n+3} + U_{n+3} + N(\mathfrak{U}) \subset U + U_{n+1}$$

for all U. Hence $x - z \in U_{n+1}$, since U_{n+1} is closed. From this follows $K(x) - y \in K(x) - K(z) + \widetilde{K(U_{n+4})} \subset K(U_{n+1}) + \widetilde{K(U_{n+4})} \subset \widetilde{K(U_n)}$.

Combining (2) and (5) gives

(6) <u>A t.v.s.</u> $E(\mathfrak{T})$ <u>is barrelled, if and only if every closed linear mapping from</u> $E(\mathfrak{T})$ <u>into an (F)-space is continuous.</u>

This is a generalization of BANACH's classical closed graph theorem. For the spaces $E(\mathfrak{T}_1)$ we need exactly the barrelled spaces, a larger class than the class of (F)-spaces.

Now we will try to find a more general class of spaces $F(\mathfrak{T}_2)$, i.e. we ask for the class of spaces $F(\mathfrak{T}_2)$, such that every closed linear mapping from a barrelled space into $F(\mathfrak{T}_2)$ is continuous. The proof of (2) shows how to find this class.

Let A be a closed linear mapping from a barrelled space $E(\mathfrak{T}_1)$ into a t.v.s. $F(\mathfrak{T}_2)$. By (3) there is a Hausdorff linear topology $\mathfrak{T}_3$ on F, which is coarser than $\mathfrak{T}_2$, such that $A: E(\mathfrak{T}_1) \longrightarrow F(\mathfrak{T}_3)$ is continuous. Then A is also a continuous mapping from $E(\mathfrak{T}_1^t) = E(\mathfrak{T}_1)$ into $F(\mathfrak{T}_3^t)$ by 6.(13). If we have $\mathfrak{T}_2 \subset \mathfrak{T}_3^t$, then $A: E(\mathfrak{T}_1) \longrightarrow F(\mathfrak{T}_2)$ will be continuous. This motivates the following definition:

A t.v.s. $F(\mathfrak{T}_0)$ is called *infra-s-space*, if for every coarser Hausdorff linear topology $\mathfrak{T}$ on F we have $\mathfrak{T}^t \supset \mathfrak{T}_0$ or, equivalently, $\mathfrak{T}^t = \mathfrak{T}_0^t$.

For these spaces the following closed graph theorem holds:

(7) *Every closed linear mapping* A *from a barrelled space* $E(\mathfrak{T}_1)$ *into an infra-s-space* $F(\mathfrak{T}_2)$ *is continuous.*

We will show now, that the infra-s-spaces are characterized by (7).

(8) *Let* $F(\mathfrak{T}_2)$ *be a t.v.s. with the property, that every closed linear mapping* A *from a barrelled space* $E(\mathfrak{T}_1)$ *into* $F(\mathfrak{T}_2)$ *is continuous. Then* $F(\mathfrak{T}_2)$ *is an infra-s-space.*

Proof. Let $\mathfrak{T}$ be a Hausdorff linear topology on F, $\mathfrak{T}$ coarser than $\mathfrak{T}_2$. Then the identity mapping I from $F(\mathfrak{T})$ onto $F(\mathfrak{T}_2)$ is closed. I remains closed considered as a mapping from $F(\mathfrak{T}^t)$ onto $F(\mathfrak{T}_2)$, since $\mathfrak{T}^t$ is finer than $\mathfrak{T}$. By hypothesis I: $F(\mathfrak{T}^t) \longrightarrow F(\mathfrak{T}_2)$ is continuous, that means $\mathfrak{T}^t \supset \mathfrak{T}_2$. With $\mathfrak{T}^t \subset \mathfrak{T}_2^t$ follows $\mathfrak{T}^t = \mathfrak{T}_2^t$ and $F(\mathfrak{T}_2)$ is an infra-s-space.

Combining (7) and (8) we obtain

(9) *A t.v.s.* $F(\mathfrak{T}_2)$ *is an infra-s-space, if and only if every closed linear mapping* A *from a barrelled space* $E(\mathfrak{T}_1)$ *into* $F(\mathfrak{T}_2)$ *is continuous.*

The proof of (8) shows, that the following stronger characterization of infra-s-spaces holds:

(10) *A t.v.s.* $F(\mathfrak{T}_2)$ *is an infra-s-space, if and only if every closed and one-to-one linear mapping* A *from a barrelled space* $E(\mathfrak{T}_1)$ *onto* $F(\mathfrak{T}_2)$ *is continuous.*

(9) and (2) give us the first and most important examples of infra-s-spaces:

(11) <u>Every (F)-space is an infra-s-space.</u>

References: N. ADASCH [4].

§ 9 Barrelled spaces and the open mapping theorem

Let A be a linear mapping from the t.v.s. $E(\mathcal{T}_1)$ into the t.v.s. $F(\mathcal{T}_2)$. With N(A) we denote the kernel of A, and in this section $\mathfrak{U}$ resp. $\mathfrak{V}$ stands for the set of all neighbourhoods of 0 in $E(\mathcal{T}_1)$ resp. $F(\mathcal{T}_2)$.

The mapping A is called o p e n , if it maps open sets of $E(\mathcal{T}_1)$ onto open sets of $R(A) := A(E)$, R(A) endowed with the topology $\hat{\mathcal{T}}_2$ induced by $\mathcal{T}_2$, or, equivalently, if A maps $\mathcal{T}_1$-neighbourhoods of 0 onto $\hat{\mathcal{T}}_2$-neighbourhoods of 0.

First we show (G. KÖTHE [3])

(1) If A is open, then $\bigcap_{V\in\mathfrak{V}} \overline{A^{-1}(V)} = \overline{N(A)}$.

Proof. It is obvious that the intersection is closed and contains N(A), hence it contains $\overline{N(A)}$. Conversely:

$$\begin{aligned}\bigcap_{V\in\mathfrak{V}} \overline{A^{-1}(V)} &= \bigcap_{\substack{V\in\mathfrak{V}\\ U\in\mathfrak{U}}} (A^{-1}(V)+U)\\ &\subset \bigcap_{U\in\mathfrak{U}} (A^{-1}(A(U))+U)\\ &= \bigcap_{U\in\mathfrak{U}} (U+N(A)+U)\\ &= \overline{N(A)}\ .\end{aligned}$$

Set $K(A) := \bigcap_{V\in\mathfrak{V}} \overline{A^{-1}(V)}$. We call A w e a k l y s i n g u l a r , if $K(A) = \overline{N(A)}$. The linear subspace $S(A) := \bigcap_{U\in\mathfrak{U}} \overline{A(U)}$ of $F(\mathcal{T}_2)$ is called the s i n g u l a r i t y o f A (G. KÖTHE [3]).

With this terminology (1) says that every open mapping is weakly singular. Further examples are given by:

(2) Every closed and hence every continuous linear mapping A is weakly singular.

Proof. By 8.(3) we have $\bigcap_{\substack{U\in\mathfrak{U}\\ V\in\mathfrak{V}}} (A(U)+V) = \{0\}$. Hence $N(A) = A^{-1}(\bigcap_{U,V} (A(U)+V)) = \bigcap_{U,V} (U+N(A)+A^{-1}(V)) = \bigcap_{U,V} (U+A^{-1}(V)) = K(A)$.

If K_A denotes the quotient mapping from $F(\mathfrak{T}_2)$ onto $F/S(A)(\hat{\mathfrak{T}}_2)$, the mapping $K_A \circ A$ is called the r e g u l a r c o n t r a c t i o n o f A . The regular contraction is always closed. Now we show (G. KÖTHE [3])

(3) A is open, if and only if A is weakly singular and the regular contraction of A is open.

Proof. Let A be open. Then A is weakly singular by (1). To show that $K_A \circ A$ is open, choose U and U_1 as neighbourhoods of 0 in $E(\mathfrak{T}_1)$, such that $U_1 + U_1 \subset U$. There exists a 0-neighbourhood V in $F(\mathfrak{T}_2)$ with $A(U_1) \supset V \cap R(A)$. For a 0-neighbourhood V_1 in $F(\mathfrak{T}_2)$ with $V_1 + V_1 \subset V$ we have

$$\begin{aligned} K_A(V_1) \cap K_A(R(A)) &= K_A((V_1+S(A)) \cap R(A)) \subset K_A((A(U_1)+V) \cap R(A)) \\ &\subset K_A(A(U_1) + (V \cap R(A))) \subset K_A A(U_1) + K_A A(U_1) \\ &\subset K_A A(U) . \end{aligned}$$

Conversely: Since A is weakly singular, we have $S(A) \cap R(A) = A(K(A)) = A(\overline{N(A)}) = \bigcap_{U\in\mathfrak{U}} A(U)$. For a 0-neighbourhood U in $E(\mathfrak{T}_1)$ choose a 0-neighbourhood U_1 with $U_1 + U_1 \subset U$. $K_A \circ A$ is open, therefore there exists a 0-neighbourhood V in $F(\mathfrak{T}_2)$ with $K_A A(U_1) \supset K_A(V) \cap K_A(R(A)) \supset K_A(V \cap R(A))$. Now $V \cap R(A) \subset (S(A) \cap R(A)) + A(U_1) \subset A(U_1) + A(U_1) \subset A(U)$.

A connection between the closed graph theorem and the open mapping theorem is given by (3). Assume A is a weakly singular surjective linear mapping and consider the following diagram:

$$K_A \circ A: \quad E(\tau_1) \xrightarrow{A} F(\tau_2) \xrightarrow{K_A} F/S(A) \ (\hat{\tau}_2)$$

$$E(\tau_1) \searrow \qquad E/N(K_A \circ A) \ (\hat{\tau}_1) \qquad \nearrow^{\widehat{K_A \circ A}} F/S(A)$$

$K_A \circ A$ is closed and therefore $\widehat{K_A \circ A}$ is closed. The inverse mapping $\widehat{K_A \circ A}^{-1}$ is continuous (that means $K_A \circ A$ is open), if $F/S(A)(\hat{\tau}_2)$ is barrelled and $E/N(K_A \circ A)$ $(\hat{\tau}_1)$ is an infra-s-space. In this case A would be open by (3).

A t.v.s. $E(\tau)$ is called s - s p a c e , if every quotient space $E/H(\hat{\tau})$, H a closed subspace of $E(\tau)$, is an infra-s-space.
Especially each s-space is an infra-s-space. It is an open question whether there exists an infra-s-space, which is not an s-space.
With this definition we obtain the following open mapping theorem:

(4) Every weakly singular linear mapping A from an s-space $E(\tau_1)$ onto a barrelled space $F(\tau_2)$ is open.

Since every quotient space of an (F)-space is again an (F)-space, we have with 8.(11)

(5) Every (F)-space is an s-space.

Hence the open mapping theorem (4) is a strong generalization of the Banach-Schauder theorem: Every continuous linear mapping from an (F)-space onto an (F)-space is open.

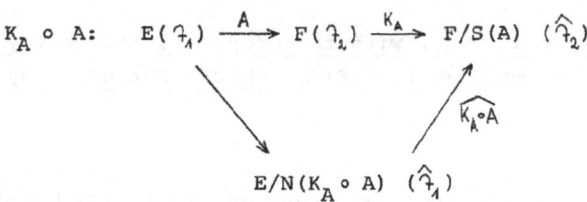

The following proposition shows that the s-spaces are characterized by (4).

(6) Let $E(\mathfrak{T}_1)$ be a t.v.s. with the property, that every weakly singular linear mapping from $E(\mathfrak{T}_1)$ onto a barrelled space $F(\mathfrak{T}_2)$ is open. Then $E(\mathfrak{T}_1)$ is an s-space.

Proof. We have to show that every quotient $E/H(\hat{\mathfrak{T}}_1)$ of $E(\mathfrak{T}_1)$ is an infra-s-space. By 8.(10) it suffices to show that every one-to-one closed linear mapping A from a barrelled space $G(\mathfrak{T})$ onto the space $E/H(\hat{\mathfrak{T}}_1)$ is continuous:

$$G(\mathfrak{T}) \xrightarrow{A} E/H(\hat{\mathfrak{T}}_1) \xleftarrow{K} E(\mathfrak{T}_1) .$$

Since with A also A^{-1} and $A^{-1} \circ K$ are closed, $A^{-1} \circ K$ is by (2) a weakly singular linear mapping from $E(\mathfrak{T}_1)$ onto the barrelled space $G(\mathfrak{T})$. By assumption on $E(\mathfrak{T}_1)$ $A^{-1} \circ K$ and therefore A^{-1} is open. Hence A is continuous.

Combining (4) and (6) we obtain

(7) A t.v.s. $E(\mathfrak{T}_1)$ is an s-space, if and only if every weakly singular linear mapping A from $E(\mathfrak{T}_1)$ onto a barrelled space $F(\mathfrak{T}_2)$ is open.

The proof of (6) shows, that it suffices to have a closed instead of a weakly singular mapping A in (6). Hence we get the following characterization of s-spaces:

(8) A t.v.s. $E(\mathfrak{T}_1)$ is an s-space, if and only if every closed linear mapping A from $E(\mathfrak{T}_1)$ onto a barrelled space is open.

References: N. ADASCH [4], G. KÖTHE [3]. For other types of closed graph and open mapping theorems see W. ROBERTSON [2].

§ 10 Completeness and the closed graph theorem

In § 8 (resp. § 9) we assumed in the closed graph theorem (resp. in the open mapping theorem), that $E(\mathfrak{T}_1)$ (resp. $F(\mathfrak{T}_2)$) is barrelled. If we replace the barrelledness of $E(\mathfrak{T}_1)$ (resp. $F(\mathfrak{T}_2)$) by " nearly continuity" of A (resp. "nearly openess" of A), we obtain the theory of B_r- (resp. B-) complete spaces. In this section we give a short outline of this theory.

A linear mapping A from a t.v.s. $E(\mathfrak{T}_1)$ into a t.v.s. $F(\mathfrak{T}_2)$ is called *nearly continuous*, if for every 0-neighbourhood V in $F(\mathfrak{T}_2)$ the set $\overline{A^{-1}(V)}$ is a 0-neighbourhood in $E(\mathfrak{T}_1)$. If A is surjective, we call A *nearly open*, if for every 0-neighbourhood U in $E(\mathfrak{T}_1)$ the set $\overline{A(U)}$ is again a 0-neighbourhood in $F(\mathfrak{T}_2)$.

With this notation we can give the following definitions: A t.v.s. $F(\mathfrak{T}_2)$ is called B_r *- complete*, if every nearly continuous closed linear mapping A from an arbitrary t.v.s. $E(\mathfrak{T}_1)$ into $F(\mathfrak{T}_2)$ is continuous.

A t.v.s. $E(\mathfrak{T}_1)$ is called *B - complete*, if every weakly singular, nearly open linear mapping A from $E(\mathfrak{T}_1)$ onto an arbitrary t.v.s. $F(\mathfrak{T}_2)$ is open.

First we show

(1) <u>A t.v.s. $F(\mathfrak{T}_2)$ is B_r-complete, if and only if every one-to-one closed, nearly continuous linear mapping A from a t.v.s. $E(\mathfrak{T}_1)$ onto $F(\mathfrak{T}_2)$ is continuous.</u>

<u>Proof.</u> Let A be a closed, nearly continuous linear mapping from a t.v.s. $E(\mathfrak{T}_1)$ into $F(\mathfrak{T}_2)$. To show that $F(\mathfrak{T}_2)$ is B_r-complete, we have to prove that A is continuous.

Let $\mathfrak{T}$ be the Hausdorff linear topology on F with the 0-neighbourhoods $A(U) + V$, where U is a 0-neighbourhood in $E(\mathfrak{T}_1)$ and V a 0-neighbourhood in $F(\mathfrak{T}_2)$ (see 8.(3)). Let I be the identity mapping from $F(\mathfrak{T}_2)$ onto $F(\mathfrak{T})$,

$$I \circ A: \quad E(\mathfrak{T}_1) \xrightarrow{A} F(\mathfrak{T}_2) \xrightarrow{I} F(\mathfrak{T}) .$$

A is continuous, if I^{-1} is continuous. I^{-1} is closed, since I is continuous. Now we will show that I is nearly open, i.e. that I^{-1} is nearly continuous. With the assumption on $F(\mathfrak{T}_2)$ follows from this, that I^{-1} is continuous, and therefore, that A is continuous.

I is nearly open: Let V and W be 0-neighbourhoods in $F(\mathfrak{T}_2)$ with $W + W \subset V$. By continuity of $I \circ A$ we obtain

$$A(\overline{A^{-1}(W)}) = (I \circ A)(\overline{A^{-1}(W)}) \subset \overline{W}^{\mathfrak{T}} ,$$

hence

$$A(\overline{A^{-1}(W)}) + W \subset \overline{W}^{\mathfrak{T}} + \overline{W}^{\mathfrak{T}} \subset \overline{V}^{\mathfrak{T}} = \overline{I(V)}^{\mathfrak{T}} .$$

Since A is nearly continuous, $\overline{I(V)}^{\mathfrak{T}}$ is a $\mathfrak{T}$-neighbourhood of 0.

From the proof of (1) we conclude, that a stronger version of (1) holds.

(2) <u>A t.v.s. $F(\mathfrak{T}_2)$ is B_r-complete, if and only if every one-to-one open, nearly continuous mapping from an arbitrary t.v.s. $E(\mathfrak{T}_1)$ onto $F(\mathfrak{T}_2)$ is a topological isomorphism.</u>

Since the algebraic isomorphisms mentioned in (1) or (2) give an identification of the spaces E and F, we obtain from these propositions an inner characterization of B_r-complete spaces.

To give this characterization we need one more definition: Let $\mathfrak{T}$ be a further linear topology on the t.v.s. $E(\mathfrak{T}_0)$. By $\overline{\mathfrak{T}_0}^{\mathfrak{T}}$ we denote the linear topology on E, whose 0-neighbourhoods are the $\mathfrak{T}$-closures $\overline{U}^{\mathfrak{T}}$ of the $\mathfrak{T}_0$-neighbourhoods U of 0. Then we have

(3) *For a t.v.s.* $E(\mathfrak{T}_0)$ *are equivalent:*

(i) $E(\mathfrak{T}_0)$ *is* B_r*-complete.*

(ii) *If* $\overline{\mathfrak{T}_0}^{\mathfrak{T}}$ *is Hausdorff for a linear topology* $\mathfrak{T}$ *on* E *and* $\overline{\mathfrak{T}_0}^{\mathfrak{T}} \subset \mathfrak{T}$, *then* $\mathfrak{T}_0 < \mathfrak{T}$.

(iii) *For each Hausdorff linear topology* $\mathfrak{T}$ *with* $\mathfrak{T} \subset \mathfrak{T}_0$ *and* $\overline{\mathfrak{T}_0}^{\mathfrak{T}} \subset \mathfrak{T}$ *we have* $\mathfrak{T}_0 = \mathfrak{T}$.

(3) is a direct consequence of (1) resp. (2). From $\overline{\mathfrak{T}_0}^{\mathfrak{T}} \subset \mathfrak{T}$, where $\overline{\mathfrak{T}_0}^{\mathfrak{T}}$ is Hausdorff, we conclude that the identity mapping I: $E(\mathfrak{T}) \longrightarrow E(\mathfrak{T}_0)$ is closed and nearly continuous.

Clearly every B-complete space is B_r-complete. Whether the converse is true, is an open question.

We want to give a further connection between B_r- and B-completeness.

(4) $E(\mathfrak{T})$ *is B-complete, if and only if every quotient space* $E(\mathfrak{T})/H$ *is* B_r*-complete.*

Proof. Assume $E(\mathfrak{T})$ is B-complete. Let H be a closed subspace and consider a one-to-one closed, nearly continuous mapping A from $F(\mathfrak{T}')$ onto $E(\mathfrak{T})/H$. If K is the quotient mapping from E onto E/H, then

$$A^{-1} \circ K: \ E(\mathfrak{T}) \xrightarrow{K} E(\mathfrak{T})/H \xrightarrow{A^{-1}} F(\mathfrak{T}')$$

is a closed, nearly open mapping from $E(\mathfrak{T})$ onto $F(\mathfrak{T}')$, hence $A^{-1} \circ K$ is open. This means A^{-1} is open and A is continuous. By (1) $E(\mathfrak{T})/H$ is B_r-complete.

Conversely, assume that every quotient space of $E(\mathfrak{T})$ is B_r-complete. For a weakly singular, nearly open mapping $A: E(\mathfrak{T}) \longrightarrow F(\mathfrak{T}')$ from E onto F we have to show, that A is open. Consider (see § 9)

$$K_A \circ A: \ E(\mathfrak{T}) \xrightarrow{A} F(\mathfrak{T}') \xrightarrow{K_A} F(\mathfrak{T}')/S(A)$$

$$\searrow \qquad\qquad \nearrow \widehat{K_A \circ A}$$

$$E(\mathfrak{T})/N(K_A \circ A)$$

$\widehat{K_A \circ A}^{-1}$ is closed and nearly continuous. Since $E(\mathfrak{T})/N(K_A \circ A)$ is B_r-complete, $\widehat{K_A \circ A}^{-1}$ is continuous and $\widehat{K_A \circ A}$ is open. Hence $K_A \circ A$ is open, and A is open by 9.(3).

Since a linear mapping from a barrelled space (onto a barrelled space) is nearly continuous (nearly open), the following proposition is obvious by the definition of B_r- (B-) completeness and 8.(9) (9.(7)).

(5) Every B_r-complete (B-complete) t.v.s. is an infra-s-space (s-space).

The converse of (5) is not true in general. However we have

(6) A barrelled infra-s-space (s-space) $E(\mathfrak{T}_0)$ is B_r-complete (B-complete).

Proof. By (4) it suffices to show (6) for infra-s-spaces. For this we use (3): Let $\mathfrak{T} \subset \mathfrak{T}_0$ be a Hausdorff linear topology on E such that $\overline{\mathfrak{T}_0}^{\mathfrak{T}} \subset \mathfrak{T}$. With $\mathfrak{T}_0$ is $\overline{\mathfrak{T}_0}^{\mathfrak{T}}$ barrelled. Since $E(\mathfrak{T}_0)$ is an infra-s-space, we have $\overline{\mathfrak{T}_0}^{\mathfrak{T}} = (\overline{\mathfrak{T}_0}^{\mathfrak{T}})^t = \mathfrak{T}_0^t = \mathfrak{T}_0$. Hence we have $\mathfrak{T}_0 = \mathfrak{T}$, $E(\mathfrak{T}_0)$ is B_r-complete.

Especially we conclude from (6) and 9.(5)

(7) Every (F)-space is B-complete.

Now we want to show that every B_r-complete and hence every B-complete t.v.s. is complete. For this purpose we first need a lemma:

(8) Let $E(\mathcal{T}_0)$ be an incomplete t.v.s. and assume $x_0 \in \widetilde{E(\mathcal{T}_0)} \setminus E$. Then the sets

$$\{([x_0] + \widetilde{U}) \cap E\ ,\ U \in \mathfrak{U}(\mathcal{T}_0)\}\ ^{*)}$$

form a base of neighbourhoods of 0 for a Hausdorff linear topology $(\mathcal{T}_0)_1$ on E with the following properties:

a) $(\mathcal{T}_0)_1 \subset \mathcal{T}_0$, $(\mathcal{T}_0)_1 \neq \mathcal{T}_0$,

b) $\overline{\mathcal{T}_0}^{(\mathcal{T}_0)_1}$ is a Hausdorff linear topology,

c) $\overline{\mathcal{T}_0}^{(\mathcal{T}_0)_1} \subset (\mathcal{T}_0)_1$.

Proof. That $(\mathcal{T}_0)_1$ is a Hausdorff linear topology is a consequence of b) and c).

a) $(\mathcal{T}_0)_1 \subset \mathcal{T}_0$ is trivial. We show $(\mathcal{T}_0)_1 \neq \mathcal{T}_0$. Let $U_0, V_0 \in \mathfrak{U}(\mathcal{T}_0)$ be balanced, such that $(x_0 + \widetilde{U}_0) \cap \widetilde{V}_0 = \emptyset$. For every balanced $U \in \mathfrak{U}(\mathcal{T}_0)$ there is an $x = \alpha x_0 + u_0$ with $\alpha > 1$, $u_0 \in \widetilde{U}_0 \cap \widetilde{U}$, such that $x \in E$. x is not contained in V_0, for otherwise we had

$$\frac{1}{\alpha} x = x_0 + \frac{1}{\alpha} u_0 \in (x_0 + \widetilde{U}_0) \cap \widetilde{V}_0 = \emptyset .$$

Hence for all $U \in \mathfrak{U}(\mathcal{T}_0)$ we have $([x_0] + \widetilde{U}) \cap E \not\subset V_0$, that means $(\mathcal{T}_0)_1 \neq \mathcal{T}_0$.

b) This follows from

$$\begin{aligned}
\bigcap_{U \in \mathfrak{U}(\mathcal{T}_0)} \overline{U}^{(\mathcal{T}_0)_1} &= \bigcap_{U, V \in \mathfrak{U}(\mathcal{T}_0)} (U + \langle ([x_0] + \widetilde{V}) \cap E \rangle) \\
&= \bigcap_{U,V} (\langle \widetilde{U} \cap E \rangle + \langle ([x_0] + \widetilde{V}) \cap E \rangle) \\
&\subset \bigcap_{U,V} (([x_0] + \widetilde{U} + \widetilde{V}) \cap E) \\
&= \Big(\bigcap_{U \in \mathfrak{U}(\mathcal{T}_0)} ([x_0] + \widetilde{U}) \Big) \cap E \\
&= [x_0] \cap E = \{0\} .
\end{aligned}$$

c) It suffices to show that for every closed $V_0 \in \mathfrak{U}(\mathcal{T}_0)$ and balanced $W_0 \in \mathfrak{U}(\mathcal{T}_0)$ with $W_0 + W_0 \subset V_0$ the inclusion

---

*) $[x_0]$ is again the one dimensional subspace spanned by x_0, $\widetilde{U}$ the closure of U in $\widetilde{E}(\widetilde{\mathcal{T}_0})$. $\mathfrak{U}(\mathcal{T}_0)$ is the set of all 0-neighbourhoods in $E(\mathcal{T}_0)$.

$$([x_0] + \widetilde{W}_0) \cap E \subset (([x_0] + \widetilde{U}) \cap E) + V_0$$

holds for every balanced $U \in \mathfrak{U}(\mathfrak{T}_0)$.

If $x \in ([x_0] + \widetilde{W}_0) \cap E$, we have $x = \alpha x_0 + w_0 \in E$ with $w_0 \in \widetilde{W}_0$. Since E is dense in $\widetilde{E}(\widetilde{\mathfrak{T}}_0)$, there is an $x_1 = -\alpha x_0 + w_1 \in E$, $w_1 \in \widetilde{W}_0 \cap \widetilde{U}$. Hence

$$x + x_1 = w_0 + w_1 \in (\widetilde{W}_0 + (\widetilde{W}_0 \cap \widetilde{U})) \cap E ,$$

that means

$$\begin{aligned} x &\in -x_1 + \langle(\widetilde{W}_0 + (\widetilde{W}_0 \cap \widetilde{U})) \cap E\rangle \\ &\subset \langle([x_0] + (\widetilde{W}_0 \cap \widetilde{U})) \cap E\rangle + \langle(\widetilde{W}_0 + (\widetilde{W}_0 \cap \widetilde{U})) \cap E\rangle . \end{aligned}$$

Therefore

$$\begin{aligned} x &\in (([x_0] + \widetilde{U}) \cap E) + (\widetilde{V}_0 \cap E) \\ &= (([x_0] + \widetilde{U}) \cap E) + V_0 . \end{aligned}$$

This proves c).

Together with (3) we obtain from (8)

(9) <u>Every B_r-complete and therefore every B-complete t.v.s.</u> $E(\mathfrak{T}_0)$ <u>is complete.</u>

In Banach's classical closed graph theorem (E and F are (F)-spaces) both spaces are complete. Replacing E by a barrelled and F by a B_r-complete space, we have lost the completeness of E, F remains complete. Replacing F by an infra-s-space we have also lost the completeness of F. However

(10) <u>Every infra-s-space</u> $E(\mathfrak{T}_0)$ <u>is complete under its associated barrelled topology</u> $\mathfrak{T}_0^t$.

<u>Proof.</u> Assume $E(\mathfrak{T}_0^t)$ is not complete. Choose $x_0 \in \widetilde{E(\mathfrak{T}_0^t)} \setminus E$, this means $x_0 \in \widetilde{E(\mathfrak{T}_0)} \setminus E$ (see 6.(14)). Let $(\mathfrak{T}_0)_1$ and $(\mathfrak{T}_0^t)_1$ denote

the topologies on E constructed in (8) with respect to x_o, $\mathfrak{T}_0$ and $\mathfrak{T}_0^t$. Then we have

(i) a) $(\mathfrak{T}_0)_1 \subset \mathfrak{T}_0$,

b) $(\mathfrak{T}_0^t)_1 \subset \mathfrak{T}_0^t$,

c) $\overline{\mathfrak{T}_0^t}^{(\mathfrak{T}_0^t)_1} \subset (\mathfrak{T}_0^t)_1$,

d) $(\mathfrak{T}_0)_1 \subset (\mathfrak{T}_0^t)_1$,

e) $(\mathfrak{T}_0)_1 \subset \overline{\mathfrak{T}_0^t}^{(\mathfrak{T}_0^t)_1}$.

a), b), c) are direct consequences of (8), while d) follows from 6.(14). For the proof of e) let U be a closed $(\mathfrak{T}_0)_1$-neighbourhood of 0. By b) and d) U is a $(\mathfrak{T}_0^t)_1$-closed $\mathfrak{T}_0^t$-neighbourhood of 0, that means U is a $\overline{\mathfrak{T}_0^t}^{(\mathfrak{T}_0^t)_1}$-neighbourhood of 0 . Now we obtain from (i):

(ii) $(\mathfrak{T}_0)_1 \subset \overline{\mathfrak{T}_0^t}^{(\mathfrak{T}_0^t)_1} \subset (\mathfrak{T}_0^t)_1 \subset \mathfrak{T}_0^t$.

$\overline{\mathfrak{T}_0^t}^{(\mathfrak{T}_0^t)_1}$ is barrelled, since $\mathfrak{T}_0^t$ is barrelled. We conclude from (ii):

(iii) $((\mathfrak{T}_0)_1)^t \subset \overline{\mathfrak{T}_0^t}^{(\mathfrak{T}_0^t)_1} \subset (\mathfrak{T}_0^t)_1 \subset \mathfrak{T}_0^t$.

By (8) $(\mathfrak{T}_0^t)_1 \neq \mathfrak{T}_0^t$, and hence by (iii) we have $((\mathfrak{T}_0)_1)^t \neq \mathfrak{T}_0^t$, that means $E(\mathfrak{T}_0)$ is no infra-s-space.

As an application of (10) we give a closed graph theorem, which in the locally convex case was proved for B-complete spaces by A.P. and W. ROBERTSON [1].

(11) <u>If</u> $E(\mathfrak{T}) = \sum_{\alpha\in I} A_\alpha(E_\alpha(\mathfrak{T}_\alpha))$ <u>is an inductive limit of Baire t.v.s.</u> $E_\alpha(\mathfrak{T}_\alpha)$ <u>and</u> $F(\mathfrak{T}') = \sum_{k=1}^{\infty} B_k(F_k(\mathfrak{T}_k))$ <u>an inductive limit of a sequence of s-spaces</u> $F_k(\mathfrak{T}_k)$ <u>with</u> $F = \bigcup_{k=1}^{\infty} B_k(F_k)$, <u>then a closed linear mapping</u> $A: E(\mathfrak{T}) \longrightarrow F(\mathfrak{T}')$ <u>is continuous.</u>

Proof. a) First we assume, that $E(\mathfrak{T})$ itself is a Baire space and that the B_k are injective, i.e. $F_k \subset F$. Then we prove $E = A^{-1}(F_{k_0})$ for a $k_0 \in \mathbb{N}$ (and A is continuous by 6.(1) and 8.(7)):

It is $E = \bigcup_{k=1}^{\infty} A^{-1}(F_k)$ and therefore an $A^{-1}(F_{k_0})$ is of second category in $E(\mathfrak{T})$. This $A^{-1}(F_{k_0})$ is dense in $E(\mathfrak{T})$ and $A^{-1}(F_{k_0})(\hat{\mathfrak{T}})$ is barrelled.

If $A_0 := A|_{A^{-1}(F_{k_0})}$, then $A_0: A^{-1}(F_{k_0})(\hat{\mathfrak{T}}) \longrightarrow F_{k_0}(\mathfrak{T}_{k_0})$ is closed and therefore continuous by 8.(7). From 6.(13) follows the continuity of $A_0: A^{-1}(F_{k_0})(\hat{\mathfrak{T}}) \longrightarrow F_{k_0}(\mathfrak{T}_{k_0}{}^t)$.

$G(A_0)$ is also closed in $E(\mathfrak{T}) \times F_{k_0}(\mathfrak{T}_{k_0})$ and hence in $E(\mathfrak{T}) \times F_{k_0}(\mathfrak{T}_{k_0}{}^t)$. $F_{k_0}(\mathfrak{T}_{k_0}{}^t)$ is complete by (10), and since the mapping

$$E(\mathfrak{T}) \supset A^{-1}(F_{k_0}) \longrightarrow F_{k_0}(\mathfrak{T}_{k_0}{}^t)$$

is continuous and closed (in $E(\mathfrak{T}) \times F_{k_0}(\mathfrak{T}_{k_0}{}^t)$), the domain $A^{-1}(F_{k_0})$ of A_0 must be closed in $E(\mathfrak{T})$, i.e. $A^{-1}(F_{k_0}) = E$.

b) For arbitrary mappings B_k the proof can be reduced to case a), since $F(\mathfrak{T}')$ is an inductive limit of the $F_k(\mathfrak{T}_k)/N(B_k)$ with injective mappings, where these spaces are infra-s-spaces.

c) If now $E(\mathfrak{T}) = \sum_{\alpha \in I} A_\alpha(E_\alpha(\mathfrak{T}_\alpha))$ with Baire spaces $E_\alpha(\mathfrak{T}_\alpha)$, then all $A \circ A_\alpha: E_\alpha(\mathfrak{T}_\alpha) \longrightarrow F(\mathfrak{T}')$ are closed, hence continuous by b). Therefore A is continuous by 4.(4).

Especially (11) gives an interesting closed graph theorem for inductive limits of (F)-spaces.

Part a) of the last proof shows that (11) is also true for infra-s-spaces $F_k(\mathfrak{T}_k)$, if the B_k are injective mappings.

The proof also shows that for a closed mapping A from a Baire space $E(\mathfrak{T})$ into a space $F(\mathfrak{T}')$ as above we have $A(E) \subset B_k(F_k)$ for a certain k.

At last an open mapping theorem can be derived in the usual manner from (11) (see § 9):

(12) *If* $E(\mathfrak{T})$ *and* $F(\mathfrak{T}')$ *are spaces as in (11), then each closed (or, more generally, each weakly singular) linear mapping from* $F(\mathfrak{T}')$ *onto* $E(\mathfrak{T})$ *is open.*

References: N. ADASCH [4], [5].

§ 11 Bornological spaces

A linear mapping A from a t.v.s. $E(\mathfrak{T}_1)$ into another t.v.s. $F(\mathfrak{T}_2)$ is b o u n d e d , if A maps bounded sets of $E(\mathfrak{T}_1)$ into bounded sets of $F(\mathfrak{T}_2)$.

If $E(\mathfrak{T}_1)$ is metrizable, every bounded mapping on $E(\mathfrak{T}_1)$ is continuous: Let $A: E(\mathfrak{T}_1) \longrightarrow F(\mathfrak{T}_2)$ be a bounded mapping, and let $\mathfrak{U} = (U_n)$ be a string in E, which generates $\mathfrak{T}_1$. Let furthermore V be a neighbourhood of 0 in $F(\mathfrak{T}_2)$. If $A^{-1}(V)$ were no 0-neighbourhood in $E(\mathfrak{T}_1)$, there would exist a sequence (x_n) in E such that $x_n \in \frac{1}{n} U_n$, but $x_n \notin A^{-1}(V)$. The sequence (nx_n) would be bounded in $E(\mathfrak{T}_1)$, but $A(nx_n)$ would not be bounded in $F(\mathfrak{T}_2)$.

A t.v.s. $E(\mathfrak{T})$ is called b o r n o l o g i c a l (i n $\mathcal{L}$ or $\mathcal{L}$ - b o r n o l o g i c a l *)) , if every bounded linear mapping on $E(\mathfrak{T})$ is continuous. We have shown so far:

(1) Metrizable t.v.s. and, in particular, (F)-spaces are bornological.

There exist also simple examples of non bornological spaces: If (E,E') is a dual pair, the identity mapping $I: E(\mathfrak{T}_s(E')) \longrightarrow E(\mathfrak{T}_k(E'))$ is bounded, but in general I is not continuous.

For a cardinal number d let φ_d be the space of all finite vectors with d real or complex coordinates. φ_d has the same bounded sets in its finest linear topology $\mathfrak{T}^f$ and in its finest locally convex topology $\mathfrak{T}^c$ (cf. 5.(10)). Hence the identity mapping $I: \varphi_d(\mathfrak{T}^c) \longrightarrow$

---

*) See the footnote on page 31 .

$\varphi_d(\mathfrak{T}^f)$ is bounded, but not continuous, if d is uncountable (see 1.(6)).

A string $\mathfrak{U} = (U_n)$ in a t.v.s. $E(\mathfrak{T})$ is called ($\mathfrak{T}$-) b o r n i - v o r o u s , if every knot U_n absorbs all bounded sets in $E(\mathfrak{T})$.

(2) <u>For a t.v.s.</u> $E(\mathfrak{T}_1)$ <u>the following assertions are equivalent:</u>

(i) $E(\mathfrak{T}_1)$ <u>is bornological.</u>

(ii) <u>Every bounded mapping</u> $A: E(\mathfrak{T}_1) \longrightarrow F(\mathfrak{T}_2)$ <u>into any (F)-space</u> $F(\mathfrak{T}_2)$ <u>is continuous.</u>

(iii) <u>Every bornivorous string in</u> $E(\mathfrak{T}_1)$ <u>is topological.</u>

<u>Proof.</u> (i) $\Rightarrow$ (ii) is obvious. (ii) $\Rightarrow$ (iii): Let $\mathfrak{U} = (U_n)$ be a bornivorous string in $E(\mathfrak{T}_1)$. Then the quotient mapping $K_{\mathfrak{U}}: E(\mathfrak{T}_1) \longrightarrow \widehat{E}_{\mathfrak{U}}$ is bounded, hence by (ii) continuous. That means $\mathfrak{U}$ is topological. (iii) $\Rightarrow$ (i): Let $A: E(\mathfrak{T}_1) \longrightarrow G(\mathfrak{T})$ be a bounded linear mapping. For every topological string $\mathfrak{W} = (V_n)$ in $G(\mathfrak{T})$ the string $A^{-1}(\mathfrak{W})$ is bornivorous in $E(\mathfrak{T}_1)$, and by (iii) it is topological, i.e. A is continuous.

If $E(\mathfrak{T})$ is a t.v.s., the set of all bornivorous strings in $E(\mathfrak{T})$ generates a linear topology $\mathfrak{T}^\beta$ on E. $\mathfrak{T}^\beta$ is finer than $\mathfrak{T}$ and $E(\mathfrak{T})$ and $E(\mathfrak{T}^\beta)$ have the same bounded sets. $\mathfrak{T}^\beta$ is the finest linear topology on E with this property. $E(\mathfrak{T}^\beta)$ is bornological. $\mathfrak{T}^\beta$ is the coarsest bornological topology on E, which is finer than $\mathfrak{T}$. $E(\mathfrak{T}^\beta)$ is called the a s s o c i a t e d b o r n o l o g i c a l s p a - c e of $E(\mathfrak{T})$. We have

(3) <u>A t.v.s.</u> $E(\mathfrak{T})$ <u>is bornological, if and only if</u> $\mathfrak{T} = \mathfrak{T}^\beta$.

Since the inverse image of a bornivorous string under a continuous linear mapping is again bornivorous, we obtain with 4.(1):

(4) <u>If</u> $E(\mathcal{T}) = \sum_{\alpha \in I} A_\alpha(E_\alpha(\mathcal{T}_\alpha))$ <u>is the inductive limit of the bornological spaces</u> $E_\alpha(\mathcal{T}_\alpha)$ <u>with respect to</u> A_α, <u>then</u> $E(\mathcal{T})$ <u>is bornological.</u>

Especially we have

(5) <u>The topological direct sum of bornological spaces is bornological. Every quotient space of a bornological space is bornological.</u>

A result similar to (4) for projective limits does not hold. For subspaces one can show that every subspace of finite codimension of a bornological space is again bornological (cf. N. ADASCH, B. ERNST [1]).

We now want to consider products of bornological spaces. First we show

(6) <u>For a countable product</u> $E(\mathcal{T}) = \prod_{i=1}^{\infty} E_i(\mathcal{T}_i)$ <u>follows</u> $E(\mathcal{T}^\beta) = \prod_{i=1}^{\infty} E_i(\mathcal{T}_i^\beta)$.

<u>Proof.</u> If $\mathcal{T}'$ denotes the topology of $\prod_{i=1}^{\infty} E_i(\mathcal{T}_i^\beta)$, then $\mathcal{T}' \subset \mathcal{T}^\beta$. For $\mathcal{T}' = \mathcal{T}^\beta$ we have to show, that every bornivorous string $\mathfrak{U} = (U_n)$ in $E(\mathcal{T})$ is topological in $E(\mathcal{T}')$.

If $U_k \in \mathfrak{U}$, there exists an $i_0 \in \mathbb{N}$, such that $U_{k+1} \supset \prod_{i=i_0}^{\infty} E_i$. To prove this assume, there is no such i_0. Then there exists a sequence (x^n) with $x^n = (x_i^n) \in \prod_{i=1}^{\infty} E_i$, such that for each x^n we have $x_i^n = 0$ for all $i \leq n$ and $x^n \notin nU_{k+1}$. Hence U_{k+1} would not absorb the bounded sequence (x^n).

$(U_n \cap \prod_{i=1}^{i_0-1} E_i)$ is a bornivorous string in $\prod_{i=1}^{i_0-1} E_i(\mathcal{T}_i)$ and hence a topological string in $\prod_{i=1}^{i_0-1} E_i(\mathcal{T}_i^\beta)$. $U_{k+1} \cap \prod_{i=1}^{i_0-1} E_i$ is therefore a 0-

neighbourhood in $\prod_{i=1}^{i_0-1} E_i(\mathcal{T}_i^\beta)$ and $(U_{k+1} \cap \prod_{i=1}^{i_0-1} E_i) + \prod_{i=i_0}^{\infty} E_i$ is a 0-neighbourhood in $\prod_{i=1}^{\infty} E_i(\mathcal{T}_i^\beta)$. From

$$U_k \supset U_{k+1} + U_{k+1} \supset (U_{k+1} \cap \prod_{i=1}^{i_0-1} E_i) + \prod_{i=i_0}^{\infty} E_i$$

follows, that U_k is a 0-neighbourhood in $\prod_{i=1}^{\infty} E_i(\mathcal{T}_i^\beta)$.

From (6) follows

(7) <u>The product $\prod_{i=1}^{\infty} E_i(\mathcal{T}_i)$ of countably many bornological spaces $E_i(\mathcal{T}_i)$ is bornological.</u>

Wether (7) is true for products of an arbitrary number of spaces depends essentially on the behavior of the spaces ω_d (for a cardinal number d we denote by ω_d the topological product of d copies of the scalar field $\mathbb{K}$).

Let E_α, $\alpha \in I$, be vector spaces, choose $x_\alpha \in E_\alpha$ and form the topological product $\prod_{\alpha\in I}[x_\alpha]$, where $[x_\alpha]$ is endowed with its natural topology. $\prod_{\alpha\in I}[x_\alpha]$ is called a s i m p l e s u b s p a c e of $\prod_{\alpha\in I} E_\alpha$. We have

(8) <u>Let $\mathfrak{U} = (U_n)$ be a string in $\prod_{\alpha\in I} E_\alpha$, which induces a topological string on every simple subspace of $\prod_{\alpha\in I} E_\alpha$. Then there is a finite subset $e \subset I$, such that $U_1 \supset \prod_{\alpha\in I\setminus e} E_\alpha$.</u>

<u>Proof.</u> As the proof of 6.(7) shows there is a finite subset e in I, such that $U_2 \supset \bigoplus_{\alpha\in I\setminus e} E_\alpha$. If $(x_\alpha) \in \prod_{\alpha\in I\setminus e} E_\alpha$, then $U_2 \cap \prod_{\alpha\in I\setminus e}[x_\alpha]$ is by assumption a 0-neighbourhood in $\prod_{\alpha\in I\setminus e}[x_\alpha]$ with $\bigoplus_{\alpha\in I\setminus e}[x_\alpha] \subset U_2 \cap \prod_{\alpha\in I\setminus e}[x_\alpha]$, and we have

$$\prod_{\alpha\in I\setminus e}[x_\alpha] = \overline{\bigoplus_{\alpha\in I\setminus e}[x_\alpha]} \subset \bigoplus_{\alpha\in I\setminus e}[x_\alpha] + (U_2 \cap \prod_{\alpha\in I\setminus e}[x_\alpha])$$
$$\subset U_2 + U_2 \subset U_1 \quad .$$

Hence $(x_\alpha) \in U_1$.

From (8) we obtain easily (cf. M. DE WILDE [1])

(9) Let $E_\alpha(\mathcal{T}_\alpha)$, $\alpha \in I$, be t.v.s.. Let $\mathfrak{U} = (U_n)$ be a string in $\prod_{\alpha\in I} E_\alpha$, which induces a topological string on every $E_\alpha(\mathcal{T}_\alpha)$, $\alpha \in I$, and on every simple subspace of $\prod_{\alpha\in I} E_\alpha(\mathcal{T}_\alpha)$. Then $\mathfrak{U}$ is topological in $\prod_{\alpha\in I} E_\alpha(\mathcal{T}_\alpha)$.

Proof. Take $U_n \in \mathfrak{U}$. For U_{n+1} there exists by (8) a finite subset e of I with $\prod_{\alpha\in I\setminus e} E_\alpha \subset U_{n+1}$. By hypothesis $U_{n+1} \cap \prod_{\alpha\in e} E_\alpha$ is a 0-neighbourhood in $\prod_{\alpha\in e} E_\alpha(\mathcal{T}_\alpha)$. Since

$$(U_{n+1} \cap \prod_{\alpha\in e} E_\alpha) + \prod_{\alpha\in I\setminus e} E_\alpha \subset U_{n+1} + U_{n+1} \subset U_n ,$$

we have shown, that U_n is a 0-neighbourhood in $\prod_{\alpha\in I} E_\alpha(\mathcal{T}_\alpha)$.

Now we can prove

(10) For a cardinal number d the following two assertions are equivalent:

(i) The product ω_d is bornological.

(ii) Every product $\prod_{\alpha\in I} E_\alpha(\mathcal{T}_\alpha)$ of d bornological spaces $E_\alpha(\mathcal{T}_\alpha)$ is bornological.

Proof. (i) $\Rightarrow$ (ii): Every simple subspace $\prod_{\alpha\in I}[x_\alpha]$ of $\prod_{\alpha\in I} E_\alpha(\mathcal{T}_\alpha)$, $x_\alpha \in E_\alpha$ with $x_\alpha \neq 0$, is topologically isomorphic to ω_d. A bornivorous string $\mathfrak{U}$ in $\prod_{\alpha\in I} E_\alpha(\mathcal{T}_\alpha)$ induces on the simple subspaces and on every $E_\alpha(\mathcal{T}_\alpha)$ a bornivorous, hence topological string. By (9) $\mathfrak{U}$ is topological in $\prod_{\alpha\in I} E_\alpha(\mathcal{T}_\alpha)$.

(7) is a consequence of (10), since for a countable d the space $\omega_d = \omega$ is metrizable, hence bornological. However it is not known, whether ω_d is for every cardinal number d a bornological space.

Detailed considerations on the question, for which numbers d the space ω_d is bornological, can be found in H. PFISTER [1]. In this paper it is shown, that ω_d is bornological, if and only if d is a non measurable cardinal number (this is analogous to the theory of locally convex spaces). However, the continuum hypothesis is assumed for the proof of this equivalence. From this results follows also, that the locally convex space ω_d is bornological in $\mathcal{L}$, if and only if it is bornological in $\mathcal{C}$.

Similar as in the locally convex case one can prove, that every bornological space $E(\mathcal{T})$ can be written as an inductive limit of certain "elementary bornological spaces E_B" generated by the bounded subsets B of $E(\mathcal{T})$. For the construction and properties of these E_B see the end of § 17 .

References: N. ADASCH [2], S.O. IYAHEN [1], S. TOMASEK [2].

§ 12 Spaces of continuous linear mappings and their completion

Let $E(\mathfrak{T}_1)$ and $F(\mathfrak{T}_2)$ be t.v.s.. Then we denote by $L(E,F)$ the vector space of all continuous linear mappings from $E(\mathfrak{T}_1)$ into $F(\mathfrak{T}_2)$ and we consider $L = L(E,F)$ as a subspace of the vector space $\mathcal{L}(E,F)$ of all linear mappings from E into F.

If B resp. V is a subset of E resp. F, then we set

$$[B,V]: = \{A: A \in \mathcal{L}(E,F) \text{ with } A(B) \subset V\} .$$

If M is a subset of $\mathcal{L}(E,F)$, then $[B,V]_M: = [B,V] \cap M$.

It is possible to define various "topologies of uniform convergence" on $L(E,F)$. For this take a system $\mathfrak{S}$ of bounded subsets in $E(\mathfrak{T}_1)$ with the following properties:

(*) a) If $B_1, B_2 \in \mathfrak{S}$, there is a $B_3 \in \mathfrak{S}$ with $B_1 \cup B_2 \subset B_3$.
b) If $B \in \mathfrak{S}$, then $\lambda B \in \mathfrak{S}$ for all $\lambda \in \mathbb{K}$.
c) $\bigcup_{B\in\mathfrak{S}} B = E$.

Furthermore let $\mathfrak{M}$ be a base of neighbourhoods of 0 in $F(\mathfrak{T}_2)$. Then it is easy to see, that the system of sets

$$\{ [B,V]_L: B \in \mathfrak{S}, V \in \mathfrak{M} \}$$

gives a 0-neighbourhood base of a linear topology $\mathfrak{T}_{\mathfrak{S}}^L$ on $L(E,F)$. From (*),c) follows that $\mathfrak{T}_{\mathfrak{S}}^L$ is Hausdorff, and we denote by $L_{\mathfrak{S}}(E,F)$ the t.v.s. $L(E,F)(\mathfrak{T}_{\mathfrak{S}}^L)$. $\mathfrak{T}_{\mathfrak{S}}^L$ is the *topology of uniform convergence on the sets* $B \in \mathfrak{S}$.

If for instance $\mathfrak{S} = \mathfrak{F}$, $\mathfrak{F}$ the system of all finite subsets of E, then we can define with $\{[B,V]: B \in \mathfrak{F}, V \in \mathfrak{M}\}$ the *topology* $\mathfrak{T}_{\mathfrak{F}}$ *of pointwise convergence* on $\mathcal{L}(E,F)$, and $\mathfrak{T}_{\mathfrak{S}}^{L}$ is the topology induced by $\mathfrak{T}_{\mathfrak{S}}$ on $L(E,F)$. Hence $L_{\mathfrak{S}}(E,F)$ is a topological subspace of $\mathcal{L}_{\mathfrak{S}}(E,F) := \mathcal{L}(E,F)(\mathfrak{T}_{\mathfrak{S}})$.

We will show now, that also for arbitrary systems σ the space $L_{\sigma}(E,F)$ can be considered as a topological subspace of a t.v.s. of linear mappings from E into F, which is greater in general.

For this let $\mathfrak{B}_{\sigma} = \mathfrak{B}_{\sigma}(E,F)$ be the subspace of $\mathcal{L}(E,F)$, which consists of all linear mappings bounded on every $B \in \sigma$. For $\mathfrak{B}_{\sigma}(E,F)$ we have:

(1) <u>The sets</u> $[B,V]_{\mathfrak{B}_{\sigma}}$, <u>where</u> $B \in \sigma$, $V \in \mathfrak{M}$, <u>generate a Hausdorff linear topology</u> $\mathfrak{T}_{\sigma}$ <u>on</u> $\mathfrak{B}_{\sigma}(E,F)$.

$\mathfrak{B}_{\sigma}(E,F)$ <u>is the greatest of those linear subspaces</u> L' <u>of</u> $\mathcal{L}(E,F)$, <u>on which</u> $\{[B,V]_{L'} : B \in \sigma, V \in \mathfrak{M}\}$ <u>generates a linear topology.</u>

<u>Proof.</u> That the $[B,V]_{\mathfrak{B}_{\sigma}}$ generate a linear topology follows, since each $[B,V]_{\mathfrak{B}_{\sigma}}$ is absorbing in $\mathfrak{B}_{\sigma}(E,F)$. But $\mathfrak{B}_{\sigma}(E,F)$ is the greatest subspace of $\mathcal{L}(E,F)$, whose elements are absorbed by all $[B,V]$. This gives the second assertion.

In the following we assume always that $\mathfrak{B}_{\sigma}(E,F)$ has the topology $\mathfrak{T}_{\sigma}$ of (1), and $[B,V]_L = [B,V] \cap L = [B,V]_{\mathfrak{B}_{\sigma}} \cap L$ implies

(2) $L_{\sigma}(E,F)$ <u>is a topological subspace of</u> $\mathfrak{B}_{\sigma}(E,F)$.

That the topology $\mathfrak{T}_{\sigma}$ on $\mathfrak{B}_{\sigma}$ is a very natural topology, that can be seen, for instance, if we ask for completeness of $\mathfrak{B}_{\sigma}(E,F)$.

(3) $\mathfrak{B}_{\sigma}(E,F)$ <u>is a complete t.v.s., if</u> $F(\mathfrak{T}_{2})$ <u>is complete.</u>

<u>Proof.</u> Assume $\{A_{\alpha}\}$ is a Cauchy system in $\mathfrak{B}_{\sigma}(E,F)$. From (*),c)

follows that $\mathfrak{T}_\sigma$ is finer than the topology of pointwise convergence, hence $\{A_\alpha\}$ is also a Cauchy system in $\mathcal{L}_s(E,F)$. Now the completeness of $F(\mathfrak{T}_2)$ implies, that there is a linear mapping A_0 with $A_\alpha \longrightarrow A_0$ in $\mathcal{L}_s(E,F)$.

But $A_0 \in \mathfrak{L}_\sigma(E,F)$: Choose $B \in \sigma$ and $V \in \mathfrak{M}$ (we assume that the elements of $\mathfrak{M}$ are closed). Since $\{A_\alpha\}$ is a $\mathfrak{T}_\sigma$-Cauchy system, there is an $\alpha_0 = \alpha_0(B,V)$ with $A_\alpha - A_\beta \in [B,V]$ for all $\alpha, \beta \geq \alpha_0$. For a closed V the set $[B,V]$ is closed in $\mathcal{L}_s(E,F)$, hence

$$(**) \qquad A_0 - A_\beta \in [B,V] \quad \text{for } \beta \geq \alpha_0 .$$

This gives especially $A_0 = A_\beta + (A_0 - A_\beta) \in \mathfrak{L}_\sigma(E,F)$ as we wanted to show.

Finally (**) implies also, that $A_\alpha \longrightarrow A_0$ in $\mathfrak{L}_\sigma(E,F)$.

With (3) we can see immediately

(4) <u>If</u> $E(\mathfrak{T}_1)$ <u>is a bornological and</u> $F(\mathfrak{T}_2)$ <u>a complete t.v.s. and if</u> $\sigma = \mathfrak{B}$ ($\mathfrak{B}$ <u>the system of all bounded sets in</u> $E(\mathfrak{T}_1)$), <u>then</u> $L_\sigma(E,F)$ ($= \mathfrak{L}_\sigma$) <u>is complete.</u>

Since the completion of a t.v.s., which is subspace of a complete t.v.s., can be obtained by the closure of this subspace, we get from (3) at once

(5) <u>If</u> $F(\mathfrak{T}_2)$ <u>is complete, then the completion</u> $\widetilde{L_\sigma(E,F)}$ <u>of</u> $L_\sigma(E,F)$ <u>has the representation</u>

$$\widetilde{L_\sigma(E,F)} = \bigcap_{\substack{B\in\sigma \\ V\in\mathfrak{M}}} \left(L(E,F) + [B,V]_{\mathfrak{L}_\sigma} \right) = \bigcap_{\substack{B\in\sigma \\ V\in\mathfrak{M}}} \left(L(E,F) + [B,V] \right) ,$$

<u>and the topology</u> $\widetilde{\mathfrak{T}}_\sigma^L$ <u>of the completion is the topology induced by</u> $\mathfrak{T}_\sigma$.

(5) gives us the following completeness criterion:

(6) Let $F(\mathfrak{T}_2)$ be complete. $L_\sigma(E,F)$ is complete, if and only if those linear mappings A from $E(\mathfrak{T}_1)$ into $F(\mathfrak{T}_2)$ are continuous, which have for each $B \in \sigma$ and $V \in \mathfrak{M}$ a decomposition $A = A_1 + A_2$ with $A_1 \in L(E,F)$ and $A_2 \in [B,V]$.

Later on we will consider some applications of (5) and (6). Finally let us remark, that the representations of the completion of a locally convex space, as they were obtained by A.P. ROBERTSON, W. ROBERTSON [2] and A. GROTHENDIECK (see G. KÖTHE [4]), and the resulting completeness criteria are special cases of (5) and (6).

References: N. ADASCH [3].

§ 13 Quasibarrelled spaces

In this section we consider an important class of t.v.s. which contains the bornological and the barrelled spaces as proper subclasses.

We call a t.v.s. $E(\mathfrak{T})$ *quasibarrelled* (in $\mathcal{L}$), if every closed bornivorous string in $E(\mathfrak{T})$ is a topological string (see S.O. IYAHEN [1]).

If $E(\mathfrak{T})$ is an arbitrary t.v.s., we denote by $\mathfrak{T}^{b^*}$ the linear topology on E which is generated by all closed and bornivorous strings in $E(\mathfrak{T})$. We have $\mathfrak{T} \subset \mathfrak{T}^{b^*}$ and the bounded sets in $E(\mathfrak{T})$ and $E(\mathfrak{T}^{b^*})$ are the same. It is obvious that $E(\mathfrak{T})$ is quasibarrelled, if and only if $\mathfrak{T} = \mathfrak{T}^{b^*}$. Further we have

(1) Every bornological and every barrelled t.v.s. is quasibarrelled.

We obtain some characterizations of quasibarrelled spaces by

(2) For a t.v.s. $E(\mathfrak{T}_1)$ the following conditions are equivalent:

(i) $E(\mathfrak{T}_1)$ is quasibarrelled.

(ii) Every bounded closed linear mapping A from $E(\mathfrak{T}_1)$ into an (F)-space $F(\mathfrak{T}_2)$ is continuous.

(iii) For every (F)-space $F(\mathfrak{T}_2)$ coincide the bounded and the equicontinuous sets in $L_b(E,F)$ *).

---

*) $L_b(E,F) := L_\sigma(E,F)$, where σ is the system of all bounded sets in $E(\mathfrak{T}_1)$, see § 12.

Proof. (i) $\longrightarrow$ (ii): Let $\mathfrak{W} = (V_n)$ be a closed topological string in $F(\mathfrak{I}_2)$ generating $\mathfrak{I}_2$. Then $(A^{-1}(V_n))$ is a bornivorous and $(\overline{A^{-1}(V_n)})$ a topological string in $E(\mathfrak{I}_1)$. This means that A is nearly continuous. Since $F(\mathfrak{I}_2)$ is B_r-complete A is continuous.

(ii) $\Longrightarrow$ (i): The proof of this is analogous to that of 8.(5). For (i) $\longleftrightarrow$ (iii) see the proof of 7.(3).

As in the case of barrelled spaces one proves, that inductive limits, products and completions of quasibarrelled spaces are quasibarrelled. In general, subspaces of quasibarrelled spaces need not be quasibarrelled, but subspaces of finite codimension are again quasibarrelled (cf. N. ADASCH, B. ERNST [1]).

For every t.v.s. $E(\mathfrak{I})$ there exists an a s s o c i a t e d q u a s i b a r r e l l e d t o p o l o g y $\mathfrak{I}^{qt}$, the coarsest quasibarrelled topology which is finer than $\mathfrak{I}$. Clearly $E(\mathfrak{I})$ is quasibarrelled, if and only if $\mathfrak{I} = \mathfrak{I}^{qt}$. We have $\mathfrak{I} \subset \mathfrak{I}^{b*} \subset \mathfrak{I}^{qt} \subset \mathfrak{I}^{\beta}$.

For a better insight in the theory of (DF)-spaces we now will consider two classes of t.v.s. which are slightly more general than the classes of quasibarrelled and barrelled spaces. Let d be an arbitrary infinite cardinal number. A t.v.s. $E(\mathfrak{I})$ is called d - q u a s i - b a r r e l l e d (in $\mathcal{L}$) (d - b a r r e l l e d (in $\mathcal{L}$)) , if every bornivorous string (every string) $\mathfrak{U} = (U_n)$, which is the intersection of d closed topological strings, is topological in $E(\mathfrak{I})$. If $d = |\mathbb{N}| = : \aleph_0$, we call $E(\mathfrak{I})$ c o u n t a b l y q u a s i b a r - r e l l e d (c o u n t a b l y b a r r e l l e d) .

In general, a d-quasibarrelled (d-barrelled) space is not quasibarrelled (barrelled). However we have

(3) <u>Assume that</u> $E(\mathfrak{I})$ <u>has a base of neighbourhoods of</u> 0 <u>of at most</u> d <u>elements. Then</u> $E(\mathfrak{I})$ <u>is quasibarrelled (barrelled), if and only if it is d-quasibarrelled (d-barrelled).</u>

In some cases the following proposition is more useful than the preceding one.

(4) <u>If the d-quasibarrelled (d-barrelled) space</u> $E(\mathcal{T})$ <u>has a dense subset of</u> d <u>elements, then</u> $E(\mathcal{T})$ <u>is quasibarrelled (barrelled).</u>

(4) is a special case of

(5) <u>If</u> $E(\mathcal{T})$ <u>is d-quasibarrelled (d-barrelled) and if</u> M <u>is a subset of</u> $E(\mathcal{T})$ <u>with</u> $0 \in M$, <u>which has a dense subset</u> N <u>of</u> d <u>elements, then</u> $\mathcal{T}$ <u>and</u> $\mathcal{T}^{b*}$ $(\mathcal{T}^{b})$ <u>induce on</u> M <u>the same neighbourhoods of</u> 0 .

<u>Proof.</u> We prove (5) for a d-quasibarrelled space $E(\mathcal{T})$. Assume $N = \{x_\alpha : \alpha \in I \text{ with } |I| = d\}$ is dense in M. Let $\mathfrak{V} = (V_n)$ be a closed bornivorous string in $E(\mathcal{T})$. We have to find a $\mathcal{T}$-neighbourhood U of 0 , such that $(U \cap M) \cap (E \setminus V_1) = \emptyset$. For this it suffices to construct an open $\mathcal{T}$-neighbourhood U of 0 , such that $(U \cap N) \cap (E \setminus V_1) = \emptyset$.

If $E \setminus V_1$ contains only finitely many elements of N, then the proof is obvious.

Assume now $N \cap (E \setminus V_1) = \{x_\alpha : \alpha \in I', \; d \geq |I'| \geq \aleph_0\}$. For every $\alpha \in I'$ there is a closed topological string (U_n^α), such that $x_\alpha \notin \overline{V_1 + U_1^\alpha}$. Set $W_n^\alpha := \overline{V_n + U_n^\alpha}$. Then (W_n^α) is a closed topological string in $E(\mathcal{T})$ and with $W_n := \bigcap_{\alpha \in I'} W_n^\alpha$ we obtain $V_n \subset W_n$, hence (W_n) is a bornivorous closed and therefore topological string in $E(\mathcal{T})$. For every open 0-neighbourhood U contained in W_1 we have the relation $U \cap N \cap (E \setminus V_1) = \emptyset$.

If $(x_\alpha)_{\alpha \in I}$ is a directed system in $E(\mathcal{T})$ and if $|I| = d$, then the balanced hull of (x_α) has a dense subset of d elements and we obtain

(6) If $E(\mathfrak{T})$ is d-quasibarrelled (d-barrelled), then a directed system $(x_\alpha)_{\alpha\in I}$ with $|I| = d$ is $\mathfrak{T}$-convergent, if and only if it is $\mathfrak{T}^{b*}$-convergent ($\mathfrak{T}^b$-convergent).

In a t.v.s. $E(\mathfrak{T})$ an associated d-quasibarrelled (associated d-barrelled) topology can be constructed for any cardinal number d in a similar way as the associated quasibarrelled (associated barrelled) topology (cf. K. NOUREDDINE, J. SCHMETS [1]).

Furthermore the d-(quasi-)barrelled spaces can be characterized by a Banach-Steinhaus property (see 7.(3)).

(7) For a t.v.s. $E(\mathfrak{T}_1)$ the following conditions are equivalent:

(i) $E(\mathfrak{T}_1)$ is d-quasibarrelled (d-barrelled).

(ii) Let $F(\mathfrak{T}_2)$ be an (F)-space. If $\mathfrak{A}$ is a bounded subset of $L_b(E,F)$ (a pointwise bounded subset of $L(E,F)$) with $\mathfrak{A} = \bigcup_{\alpha\in I} \mathfrak{A}_\alpha$, where $|I| \leq d$ and every $\mathfrak{A}_\alpha$ is equicontinuous, then $\mathfrak{A}$ is equicontinuous.

Proof. (i) $\Longrightarrow$ (ii) is obvious. (ii) $\Longrightarrow$ (i): For the d-quasibarrelled case let $\mathfrak{U} = (U_n)$ be a closed bornivorous string in $E(\mathfrak{T}_1)$, which is the intersection of d closed topological strings $\mathfrak{U}_\alpha = (U_n^\alpha)$. Then define $F: = \bigoplus_\alpha E_\alpha$ with $E_\alpha: = E$ for all α and $V_n: = \bigoplus_\alpha (U_n + U_n^\alpha)$ and continue as in the proof of 7.(2) .

At last we want to note that the permanence properties of the d-quasibarrelled and d-barrelled spaces are the same as in the case of quasibarrelled and barrelled spaces.

§ 14 Boundedly summing spaces

In the category $\mathcal{C}$ of all locally convex topological vector spaces every sequentially complete $\mathcal{C}$-quasibarrelled space is also $\mathcal{C}$-barrelled. In the category $\mathcal{L}$ this is in general not true as can be shown by a counter example of P. TURPIN [2] (cf. 19.3.). In this section we consider a class of topological vector spaces containing the locally convex spaces, for which the above mentioned proposition remains true.

If B is a bounded set in the t.v.s. $E(\mathcal{T})$, we denote by R(B) the set of all sequences $\lambda = (\lambda_n) \in \omega$, such that $\sum_{n=1}^{\infty} \lambda_n B$ $\left(:= \bigcup_{m=1}^{\infty} \sum_{n=1}^{m} \lambda_n B\right)$ is again bounded. Of course $\varphi \subset R(B)$.

We call a t.v.s. $E(\mathcal{T})$ b o u n d e d l y s u m m i n g , if for every bounded set B in $E(\mathcal{T})$ we have $R(B) \supsetneqq \varphi$ (i.e. if for every bounded set B there exists a sequence $\lambda = (\lambda_n)$, $\lambda_n \neq 0$, such that $\sum_{n=1}^{\infty} \lambda_n B$ is bounded).

In [1] P. TURPIN introduced the "galbe" G(E) of a t.v.s. $E(\mathcal{T})$. G(E) is the space of all sequences $\lambda = (\lambda_n) \in \omega$, such that for every neighbourhood U of 0 there exists a neighbourhood V of 0 with $\sum_{n=1}^{\infty} \lambda_n V \subset U$.

Obviously $G(E) = \ell^p$, if $E(\mathcal{T})$ is locally p-convex. One can also see immediately

(1) <u>It is</u> $\varphi \subset G(E) \subset R(B)$ <u>for every bounded set</u> B <u>in a t.v.s.</u> $E(\mathcal{T})$.

The following permanence properties of boundedly summing spaces are obvious.

(2) (i) Subspaces, countable products, countable projective limits and topological direct sums of boundedly summing spaces are boundedly summing.

(ii) If every bounded set of the quotient space $E(\mathcal{T})/H$ is contained in the closure of the canonical image of a bounded set in $E(\mathcal{T})$, then $E(\mathcal{T})/H$ is boundedly summing, if $E(\mathcal{T})$ is boundedly summing.

(iii) If every bounded set in the completion $\tilde{E}(\tilde{\mathcal{T}})$ is contained in the completion of a bounded set in $E(\mathcal{T})$ and if $E(\mathcal{T})$ is boundedly summing, then $\tilde{E}(\tilde{\mathcal{T}})$ is boundedly summing too.

In a special case we can prove a converse of (1).

(3) If $E(\mathcal{T})$ is a bornological t.v.s. with a fundamental sequence (B_n) of bounded sets B_n *), then $G(E) = \bigcap_{n=1}^{\infty} R(B_n)$.
$G(E) \neq \varphi$ holds, if and only if $E(\mathcal{T})$ is boundedly summing.

Proof. We have $G(E) \subset \bigcap_{n=1}^{\infty} R(B_n)$ by (1). We assume the B_n balanced with $B_n + B_n \subset B_{n+1}$. Choose $(\mu_i) \in \bigcap_{n=1}^{\infty} R(B_n)$, and let (U_n) be a topological string in $E(\mathcal{T})$. There are $\alpha_n > 0$, such that $\alpha_n(\sum_{i=1}^{\infty} \mu_i B_n) \subset U_{n+1}$. The sequence (V_j) with $V_j := \sum_{n=1}^{\infty} \alpha_{n+j-1} B_n$ is a bornivorous string in $E(\mathcal{T})$. Therefore (V_j) is topological, and from

$$\sum_{i=1}^{\infty} \mu_i V_1 = \sum_{i=1}^{\infty} \mu_i (\sum_{n=1}^{\infty} \alpha_n B_n) = \sum_{n=1}^{\infty} \alpha_n (\sum_{i=1}^{\infty} \mu_i B_n) \subset \sum_{n=1}^{\infty} U_{n+1} \subset U_1$$

follows $(\mu_i) \in G(E)$.

If $E(\mathcal{T})$ is boundedly summing, then $R(B_n) \neq \varphi$ for all n. Hence there are sequences $(\mu_i^n)_{i\in N}$, $\mu_i^n > 0$, such that $\sum_{i=1}^{\infty} \mu_i^n B_n$ is boun-

*) That means every bounded set in $E(\mathcal{T})$ is contained in some B_n.

ded. We can assume $\mu_i^n \geq \mu_i^{n+1}$ for all $i,n \in \mathbb{N}$. If $\lambda_i := \mu_i^i$, then $\sum_{i=1}^{\infty} \lambda_i B_n$ is bounded for each n. Therefore $(\lambda_i) \in \bigcap_{n=1}^{\infty} R(B_n) = G(E)$.

Of course, every locally convex space is boundedly summing, since for every absolutely convex bounded subset B we have $(\frac{1}{2^n}) \in R(B)$. However the class of boundedly summing spaces is greater than $\mathcal{C}$.

(4) Every metrizable t.v.s. $E(\mathcal{T})$ is boundedly summing. Almost convex and locally pseudoconvex spaces $E(\mathcal{T})$ are boundedly summing.

Proof. In the metrizable case $\sum_{n=1}^{\infty} \lambda_n B$ is bounded, if B is bounded with $\lambda_n B \subset U_n$, $\lambda_n \neq 0$, where (U_n) is a $\mathcal{T}$ generating string.

A t.v.s. $E(\mathcal{T})$ is almost convex, if it has a fundamental system $\mathfrak{B}$ of bounded balanced sets, such that for $B \in \mathfrak{B}$ there is a $\lambda > 0$ with $B + B \subset \lambda B$. Then we have $(\frac{1}{\lambda^n}) \in R(B)$.

$E(\mathcal{T})$ is locally pseudoconvex, if it has a base $\mathfrak{U}$ of balanced neighbourhoods of 0, such that for $U \in \mathfrak{U}$ there is a $\lambda > 0$ with $U + U \subset \lambda U$. $E(\mathcal{T})$ is then a countable projective limit of locally p-convex, hence almost convex spaces. By (2) the proposition follows.

In $\mathcal{C}$ each bornological space can be written as an inductive limit of normed spaces. For the class of boundedly summing spaces we have

(5) Every bornological boundedly summing space $E(\mathcal{T})$ is an inductive limit of metrizable spaces.

Proof. Let $\mathfrak{B}$ be the set of all bounded balanced sets in $E(\mathcal{T})$. For $B \in \mathfrak{B}$ we take the linear hull $[B]$ of B in E. Choose $(\lambda_n) \in R(B)$ with $\lambda_n > \lambda_{n+1} > 0$. Then the string (V_n) in $[B]$ with $V_n := \sum_{j=1}^{\infty} \lambda_{2^{n-1} j} B$ generates a metrizable linear topology on $[B]$, which is finer than the topology induced by $\mathfrak{T}$. We write $E[B]$ for this t.v.s.. Since each B is bounded in $E[B]$ and $E(\mathfrak{T})$ is bornological, it follows at once that $E(\mathfrak{T})$ is representable as an inductive limit of the $E[B]$, $B \in \mathfrak{B}$.

Now we can prove

(6) <u>If</u> $E(\mathfrak{T})$ <u>is sequentially complete and boundedly summing, then</u> $\mathfrak{T}^{b*} = \mathfrak{T}^{b}$.
<u>Especially: Every sequentially complete quasibarrelled boundedly summing t.v.s. is barrelled.</u>

<u>Proof.</u> Let $\mathfrak{U} = (U_n)$ be a closed string in $E(\mathfrak{T})$ and let B be a bounded balanced subset. For the embedding $I_B: E[B] \longrightarrow E(\mathfrak{T})$ denote by $\tilde{I}_B$ the continuous extension of I_B to the completion $\widetilde{E[B]}$. Since $E(\mathfrak{T})$ is sequentially complete, $\tilde{I}_B$ is a continuous mapping from $\widetilde{E[B]}$ into $E(\mathfrak{T})$. Hence $(\tilde{I}_B)^{-1}(\mathfrak{U})$ is a closed and therefore topological string in the (F)-space $\widetilde{E[B]}$. Then B is absorbed by every $(\tilde{I}_B)^{-1}(U_n)$ and hence by every U_n.

The following example shows, that an uncountable product of boundedly summing spaces is not necessarily boundedly summing (cf. W. RUESS, R. WAGNER [1]).

We consider the (F)-space $S(0,1)$ (cf. the examples in § 2). For $j \in \mathbb{N}_0$ and $k \in \mathbb{N}$ with $1 \leq k \leq 2^j$ we define

$$f_{j,k}(x) := \begin{cases} 1 & \text{for } x \in \left[\frac{k-1}{2^j}, \frac{k}{2^j}\right] \\ 0 & \text{otherwise} . \end{cases}$$

We arrange the $f_{j,k}$ in an obvious way to a sequence (f_n), such that (f_n) is a null sequence in $S(0,1)$ and this holds for every sequence $(\lambda_n f_n)$ with $\lambda_n \in \mathbb{K}$.

For the sequence space ω we consider the product $\prod_{\alpha\in\omega} S(0,1)_\alpha$ with $S(0,1)_\alpha := S(0,1)$. Set $N_\alpha := \overline{(\lambda_n f_n)}$ for $\alpha = (\lambda_n) \in \omega$. Then $N := \prod_{\alpha\in\omega} N_\alpha$ is bounded in $\prod_{\alpha\in\omega} S(0,1)_\alpha$.

Assume $\prod_{\alpha\in\omega} S(0,1)_\alpha$ is boundedly summing. Then there is a sequence (ϱ_m) with $\varrho_m > 0$, such that $\sum_{m=1}^{\infty} \varrho_m N$ is bounded in $\prod_{\alpha\in\omega} S(0,1)_\alpha$. This implies that $\sum_{m=1}^{\infty} \varrho_m N_{\alpha_0}$ is bounded in $S(0,1)_{\alpha_0} = S(0,1)$ for $\alpha_0 := (\frac{1}{\varrho_m})$. But this is impossible, since $\sum_{n=1}^{\ell} f_n \in \sum_{m=1}^{\infty} \varrho_m N_{\alpha_0}$ for all $\ell \in \mathbb{N}$.

In general quotients and inductive limits of boundedly summing spaces need not be boundedly summing. This shows a further example in W. RUESS, R. WAGNER [1].

§ 15 Locally topological spaces

Now we introduce a common generalization of bornological spaces and (DF)-spaces. Let $\mathfrak{B}$ be a system of bounded balanced sets in the t.v.s. $E(\mathfrak{T})$, such that for $B_1, B_2 \in \mathfrak{B}$ the sets $B_1 \cap B_2$ and $B_1 + B_2$ are also contained in $\mathfrak{B}$. For example all balanced bounded or precompact or compact sets in $E(\mathfrak{T})$ form such a system.

A linear mapping $A: E(\mathfrak{T}) \longrightarrow F(\mathfrak{T}')$ is called $\mathfrak{B}$ - l o c a l l y c o n t i n u o u s, if all restrictions $A|_B$ for $B \in \mathfrak{B}$ are continuous in 0 in the topology induced by $\mathfrak{T}$ on B.

$E(\mathfrak{T})$ is a $\mathfrak{B}$ - l o c a l l y t o p o l o g i c a l s p a c e, if all $\mathfrak{B}$-locally continuous mappings from $E(\mathfrak{T})$ into any t.v.s. $F(\mathfrak{T}')$ are continuous.

We call a string $\mathfrak{U} = (U_n)$ in $E(\mathfrak{T})$ a $\mathfrak{B}$ - l o c a l l y t o - p o l o g i c a l s t r i n g, if $U_n \cap B$ is a neighbourhood of 0 for the topology induced by $\mathfrak{T}$ on B for all $n \in \mathbb{N}$ and $B \in \mathfrak{B}$.

In the sequel we will only consider the system $\mathfrak{B} = \mathcal{B}$ of all balanced bounded sets of $E(\mathfrak{T})$ (in this case we speak of locally continuous mappings, locally topological spaces and strings). We do this for simplicity reasons only. Most of the following propositions hold also for arbitrary systems $\mathfrak{B}$.

Obviously we have

(1) <u>For a t.v.s. $E(\mathfrak{T})$ the following conditions are equivalent:</u>

(i) $E(\mathfrak{T})$ <u>is locally topological.</u>

(ii) <u>All locally continuous linear mappings from $E(\mathfrak{T})$ into an (F)-space are continuous.</u>

(iii) <u>All locally topological strings are topological.</u>

For (ii) $\Longrightarrow$ (iii) see the proof of 11.(2).

For a t.v.s. $E(\mathcal{T})$ let $\mathcal{T}^{lt}$ be the topology on E, which has all locally topological strings as a base of neighbourhoods of 0. By (1) $E(\mathcal{T})$ is locally topological, if and only if $\mathcal{T} = \mathcal{T}^{lt}$.

$\mathcal{T}^{lt}$ is by definition the finest linear topology on E which induces on every balanced bounded set B in $E(\mathcal{T})$ the same neighbourhoods of 0 as $\mathcal{T}$. But we have stronger

(2) *$\mathcal{T}^{lt}$ induces on every balanced bounded set B in $E(\mathcal{T})$ the same topology as $\mathcal{T}$. $\mathcal{T}^{lt}$ is the finest linear topology on E with this property.*

Proof. We have to show that for each locally topological string $\mathfrak{U} = (U_n)$ in $E(\mathcal{T})$ and each $x \in B$ the set $(x + U_1) \cap B$ is a neighbourhood of x in the topology induced by $\mathcal{T}$ on B. There is a neighbourhood U of 0 in $E(\mathcal{T})$, such that $(B + B) \cap U \subset U_1$. If $z \in (x + U) \cap B$, then $z - x \in (B + B) \cap U \subset U_1$ and $z \in (x + U_1) \cap B$, hence $(x + U) \cap B \subset (x + U_1) \cap B$.

Clearly a linear mapping A on $E(\mathcal{T})$ is locally continuous, if and only if it is continuous on $E(\mathcal{T}^{lt})$. By (2) this is the case, if and only if all $A|_B$ are continuous in the topology induced by $\mathcal{T}$ on B.

To obtain examples of locally topological spaces we need a proposition on the bounded sets in $E(\mathcal{T}^{lt})$.

(3) *Let B be a bounded balanced set in the t.v.s. $E(\mathcal{T})$ and let U be a balanced set, such that $B \cap U$ is a neighbourhood of 0 for the topology induced by $\mathcal{T}$ on B. Then U absorbs B.*

Proof. Assume B is not absorbed by U. Then there are $x_n \in \frac{1}{n} B$ with $x_n \notin U$. The x_n are contained in B and converge to 0. This contradicts $x_n \notin U$ for all $n \in \mathbb{N}$.

An easy consequence of (3) is

(4) A locally topological string in a t.v.s. $E(\mathfrak{T})$ is bornivorous, and $E(\mathfrak{T})$ and $E(\mathfrak{T}^{lt})$ have the same bounded sets.

(4) implies that $\mathfrak{T}^{lt}$ is coarser than the associated bornological topology $\mathfrak{T}^{\beta}$. Hence all bornological spaces are locally topological.

We now want to prove some permanence properties of locally topological spaces. Since by (4) a locally continuous mapping maps bounded sets onto bounded sets we have

(5) The product of two locally continuous mappings is locally continuous.

From this follows immediately

(6) An inductive limit $E(\mathfrak{T}) = \sum_{\alpha \in I} A_{\alpha}(E_{\alpha}(\mathfrak{T}_{\alpha}))$ of locally topological spaces $E_{\alpha}(\mathfrak{T}_{\alpha})$ is locally topological.
In particular: topological direct sums and quotients of locally topological spaces are locally topological.

The proof of the following proposition is analogous to that of 11.(6).

(7) A countable product $E(\mathfrak{T}) = \prod_{i=1}^{\infty} E_i(\mathfrak{T}_i)$ of locally topological spaces $E_i(\mathfrak{T}_i)$ is locally topological.

Before we can answer the corresponding question for arbitrary products of locally topological spaces, we need a lemma.

(8) <u>If</u> $\mathfrak{U} = (U_n)$ <u>is a locally topological string in</u> $E(\mathcal{T})$, <u>then</u> $N(\mathfrak{U}) \cap B$ <u>is closed in</u> B <u>for all</u> $B \in \mathcal{B}$.

<u>Proof.</u> Assume the directed system (x_α) with $x_\alpha \in N(\mathfrak{U}) \cap B$ converges to $x \in B$. Then $(x - x_\alpha)$ converges to 0 in $E(\mathcal{T})$. Since $x - x_\alpha \in B + B$ for all α and since U_{n+1} induces a 0-neighbourhood on $B + B$, we have $x - x_\alpha \in U_{n+1}$ for $\alpha \geq \alpha_0$. Since $x_\alpha \in U_{n+1}$, we have $x \in U_{n+1} + U_{n+1} \subset U_n$. Consequently $x \in N(\mathfrak{U})$.

(9) <u>An arbitrary product</u> $E(\mathcal{T}) = \prod_{\alpha \in I} E_\alpha(\mathcal{T}_\alpha)$ <u>of locally topological spaces</u> $E_\alpha(\mathcal{T}_\alpha)$ <u>is locally topological.</u>

<u>Proof.</u> The proof is similar to that of 6.(9). Let $\mathfrak{U} = (U_n)$ be a locally topological string in $E(\mathcal{T})$. By (4) and 6.(7) for all $n \in \mathbb{N}$ there is a finite subset $e_n \subset I$, such that $\bigoplus_{\alpha \in I \setminus e_n} E_\alpha \subset U_n$. Denote by I_0 the union of all the e_n. Then $\bigoplus_{\alpha \in I \setminus I_0} E_\alpha$ is a subset of $N(\mathfrak{U})$.

We want to show $N(\mathfrak{U}) \supset \prod_{\alpha \in I \setminus I_0} E_\alpha$. Assume $x = (x_\alpha) \in \prod_{\alpha \in I \setminus I_0} E_\alpha$. The set $X := \prod_{\alpha \in I \setminus I_0} [x_\alpha]_1$ is a product of balanced compact sets and hence closed, bounded and balanced in $\prod_{\alpha \in I \setminus I_0} E_\alpha(\mathcal{T}_\alpha)$. Since $N(\mathfrak{U}) \supset \bigoplus_{\alpha \in I \setminus I_0} E_\alpha$ and $X = \overline{X \cap \bigoplus_{\alpha \in I \setminus I_0} E_\alpha}^{\mathcal{T}}$, we obtain by (8) that $X \subset N(\mathfrak{U})$ and $x \in N(\mathfrak{U})$.

Now we have shown

$$U_1 \supset (U_2 \cap \prod_{\alpha \in I_0} E_\alpha) + \prod_{\alpha \in I \setminus I_0} E_\alpha .$$

Applying (7) we see that $U_2 \cap \prod_{\alpha \in I_0} E_\alpha$ is a neighbourhood of 0 in $\prod_{\alpha \in I_0} E_\alpha(\mathcal{T}_\alpha)$, and hence U_1 is a neighbourhood of 0 in $E(\mathcal{T})$.

We finish this section with a proposition on spaces of linear mappings.

(10) <u>Let</u> $E(\mathcal{T}_1)$ <u>be a locally topological space. Then for all complete t.v.s.</u> $F(\mathcal{T}_2)$ <u>the space</u> $L_b(E,F)$ <u>is complete.</u>

<u>Proof.</u> The proof is an easy application of the results in § 12 . Assume A is an element of the completion of $L_b(E,F)$. Let $\mathfrak{W} = (V_n)$ be a topological string in $F(\mathfrak{T}_2)$ and let B be a bounded set in $E(\mathfrak{T}_1)$. Then there are linear mappings $A_1 \in L(E,F)$ and A_2 , such that $A_2(B) \subset V_2$ and $A = A_1 + A_2$. Since A_1 is continuous, there is a topological string $\mathfrak{U} = (U_n)$ in $E(\mathfrak{T}_1)$, such that $A_1(U_1) \subset V_2$ and we have $A(B \cap U_1) = A_1(B \cap U_1) + A_2(B \cap U_1) \subset V_2 + V_2 \subset V_1$. That means A is locally continuous. Since $E(\mathfrak{T}_1)$ is locally topological, A is continuous and hence $A \in L(E,F)$.

References: N. ADASCH [7], N. ADASCH, B. ERNST [3].

§ 16 Spaces with an absorbing sequence

Before we give further results on locally topological spaces, we consider another class of spaces which will turn out as a generalization of the class of (DF)-spaces.

For this let $\sigma := (B_n)$ be a sequence of balanced sets in the vector space E, such that $B_n + B_n \subset B_{n+1}$ for all $n \in \mathbb{N}$ (to simplify our notation we will sometimes set $B_n := \{0\}$ for $n \leq 0$). σ is called *absorbing*, if $\bigcup\limits_{n=1}^{\infty} B_n$ is absorbing in E, that means if $E = \bigcup\limits_{n=1}^{\infty} B_n$.

If $E(\mathfrak{T})$ is a t.v.s., we call the sequence σ *bornivorous*, if every bounded set in $E(\mathfrak{T})$ is contained in some B_n. If furthermore all B_n are bounded, we call σ a *fundamental sequence of bounded sets* in $E(\mathfrak{T})$.

For an absorbing sequence σ in a t.v.s. $E(\mathfrak{T})$ we denote by $\mathfrak{T}_\sigma$ the finest of the linear topologies $\mathfrak{T}'$ on E, such that the embeddings of the B_n , endowed with the topology induced by $\mathfrak{T}$, into the space $E(\mathfrak{T}')$ are continuous at 0. Hence $\mathfrak{T}_\sigma$ is the finest linear topology on E, which on every B_n induces the same neighbourhoods of 0 as $\mathfrak{T}$. It is obvious that $E(\mathfrak{T}_\sigma)$ is something like a strict inductive limit.

In the same way as in the last section we obtain the 0-neighbourhoods of $\mathfrak{T}_\sigma$:

(1) <u>Let $\sigma = (B_n)$ be an absorbing sequence in the t.v.s. $E(\mathfrak{T})$. A string (U_j) is $\mathfrak{T}_\sigma$-topological, if and only if $U_j \cap B_n$ is a neighbourhood of 0 in B_n with the topology induced by $\mathfrak{T}$ for all $j, n \in \mathbb{N}$.</u>

With the proof of 15.(2) follows, that $\mathfrak{T}_\sigma$ is the finest linear topology on E which induces on every B_n the same topology as $\mathfrak{T}$. As in 15.(1) we have

(2) *If a t.v.s. $E(\mathfrak{T})$ has an absorbing sequence $\sigma = (B_n)$, then the following conditions are equivalent:*

(i) It is $\mathfrak{T} = \mathfrak{T}_\sigma$.

(ii) A linear mapping A from $E(\mathfrak{T})$ into any t.v.s. is continuous, if $A|_{B_n}$ is continuous (at 0) for every $n \in \mathbb{N}$.

(iii) A linear mapping A from $E(\mathfrak{T})$ into any (F)-space is continuous, if $A|_{B_n}$ is continuous (at 0) for every $n \in \mathbb{N}$.

A connection between the 0-neighbourhoods of $\mathfrak{T}$ and $\mathfrak{T}_\sigma$ is given by the following proposition.

(3) *Let $\sigma = (B_n)$ be an absorbing sequence in the t.v.s. $E(\mathfrak{T})$. Let $(\mathfrak{U}^n) = ((U_j^n))$ run through all sequences of topological strings in $E(\mathfrak{T})$. Then the strings*

(i) (V_j) *with* $V_j := \sum_{n=1}^{\infty} (B_{n-j+1} \cap U_j^n)$

generate the topology $\mathfrak{T}_\sigma$. The same is true for the following sets of strings:

(ii) (W_j) *with* $W_j := \bigcap_{n=1}^{\infty} (B_{n-j+1} + U_j^n)$, *or*

(iii) (W_j') *with* $W_j' := \bigcap_{n=1}^{\infty} (\overline{B_{n-j+1} + U_j^n}^{\mathfrak{T}})$, *or*

(iv) (W_j'') *with* $W_j'' := \bigcap_{n=1}^{\infty} (\overline{B_{n-j+1}}^{\mathfrak{T}} + U_j^n)$.

Proof. a) Every $\mathfrak{T}_\sigma$-topological string (S_j) contains a (V_j): For each $n \in \mathbb{N}$ there is a $\mathfrak{T}$-topological string $\mathfrak{U}^n = (U_j^n)$, such that $B_n \cap U_j^n \subset S_{n+j}$ for all $j \in \mathbb{N}$. Then $V_j := \sum_{n=1}^{\infty} (B_{n-j+1} \cap U_j^n) \subset \sum_{n=1}^{\infty} (B_n \cap U_j^n) \subset \sum_{n=1}^{\infty} S_{n+j} \subset S_j$.

b) For every (V_j) there is a (W_j), such that $V_j \supset W_{j+1}$: For (V_j) with $V_j = \sum_{n=1}^{\infty}(B_{n-j+1} \cap U_j^n)$ we can assume, that the sequence $(\mathfrak{U}^n)$ of topological strings $\mathfrak{U}^n = (U_j^n)$ is decreasing. Then choose (W_j) with $W_j := \bigcap_{n=1}^{\infty}(B_{n-j+1} + U_j^{n+1})$. If $x \in W_{j+1}$, then $x = y_n + u_n$ with $y_n \in B_{n-j}$, $u_n \in U_{j+1}^{n+1}$ for all $n \in \mathbb{N}$, and $x = \sum_{i=1}^{m} x_i + u_n$ with $x_1 := y_1$, $x_i := y_i - y_{i-1}$ for $i \geq 2$. But $x_i = y_i - y_{i-1} \in B_{i-j+1}$, on the other hand $x_i = u_{i-1} - u_i \in U_j^i$, hence $x_i \in B_{i-j+1} \cap U_j^i$. If $x \in B_{n_0-j}$ for an $n_0 \in \mathbb{N}$, then $u_{n_0} = x - y_{n_0} \in B_{n_0-j+1} \cap U_{j+1}^{n_0+1}$. Now

$$x = \sum_{i=1}^{n_0} x_i + u_{n_0} \in \sum_{i=1}^{n_0}(B_{i-j+1} \cap U_j^i) + (B_{n_0-j+1} \cap U_{j+1}^{n_0+1})$$

$$\subset V_j \quad .$$

c) Since $\overline{B_{n-j+1} + U_{j+1}^n}^{\mathfrak{T}} \subset B_{n-j+1} + U_j^n$, it is obvious that each (W_j) contains a (W_j'). In a similar way follows, that each (W_j') contains a (W_j'').

d) Finally it remains to prove, that every (W_j'') is a $\mathfrak{T}_\sigma$-topological string. Choose $m \in \mathbb{N}$. Then for $n \geq m+j$ we have $B_m \subset \overline{B_{n-j+1}}^{\mathfrak{T}} + U_j^n$. Hence

$$W_j'' \cap B_m \supset B_m \cap \bigcap_{n=1}^{m+j}(\overline{B_{n-j+1}}^{\mathfrak{T}} + U_j^n) \quad ,$$

that means W_j'' induces on B_m a $\mathfrak{T}$-neighbourhood of 0.

From (3),(iii) we obtain immediately

(4) If σ is an absorbing sequence in the t.v.s. $E(\mathfrak{T})$, then $\mathfrak{T}_\sigma$ has a base of neighbourhoods of 0 of $\mathfrak{T}$-closed sets. Hence, if $E(\mathfrak{T})$ is complete, then $E(\mathfrak{T}_\sigma)$ is complete.

(3),(iv) gives

(5) <u>If $\sigma = (B_n)$ is an absorbing sequence in</u> $E(\mathfrak{T})$ <u>and if $\overline{\sigma}$ denotes the absorbing sequence formed by the $\mathfrak{T}$-closures of the B_n, then $\mathfrak{T}_\sigma = \mathfrak{T}_{\overline{\sigma}}$.</u>

With (3) it is also easy to prove the following generalization of 5.(9).

(6) <u>If $\sigma = (B_n)$ is an absorbing sequence of closed sets in the t.v.s.</u> $E(\mathfrak{T})$, <u>then σ is bornivorous in</u> $E(\mathfrak{T}_\sigma)$.

<u>Proof.</u> Assume there is a bounded set B in $E(\mathfrak{T}_\sigma)$ which is contained in no B_n, that means there is a $\mathfrak{T}_\sigma$-0-sequence (x_n) with $x_n \in \frac{1}{n} B$ and $x_n \notin B_n$. Since the B_n are closed, there are $\mathfrak{T}$-topological strings $\mathfrak{U}^n = (U_j^n)$, such that $x_n \notin B_n + U_1^n$. By (3) the sets $W_j = \bigcap_{n=1}^{\infty} (B_{n-j+1} + U_j^n)$ form a $\mathfrak{T}_\sigma$-topological string, but W_1 contains no element of the 0-sequence (x_n).

In the following proposition we prove a condition for $\mathfrak{T}$ and $\mathfrak{T}_\sigma$ to be equal.

(7) <u>Let $\sigma = (B_n)$ be an absorbing (bornivorous) sequence in</u> $E(\mathfrak{T})$. <u>If</u> $E(\mathfrak{T})$ <u>is countably barrelled (countably quasibarrelled), then it is $\mathfrak{T} = \mathfrak{T}_\sigma$.</u>

<u>Proof.</u> By (3),(iii) we have a $\mathfrak{T}_\sigma$ generating set of strings (of bornivorous strings), which are the intersection of sequences of $\mathfrak{T}$-closed, $\mathfrak{T}$-topological strings. By assumption follows $\mathfrak{T}_\sigma \subset \mathfrak{T}$. Since always $\mathfrak{T} \subset \mathfrak{T}_\sigma$, we have $\mathfrak{T} = \mathfrak{T}_\sigma$.

If $\mathfrak{T} = \mathfrak{T}_\sigma$ for a t.v.s. $E(\mathfrak{T})$ with an absorbing sequence σ, then $E(\mathfrak{T})$ is a strict inductive limit of certain subspaces of $E(\mathfrak{T})$:

(8) <u>If</u> $\sigma = (B_n)$ <u>is an absorbing sequence in the t.v.s.</u> $E(\mathfrak{T})$ <u>with</u> $\mathfrak{T} = \mathfrak{T}_\sigma$, <u>then</u> $E(\mathfrak{T}) = \bigcup_{n=1}^{\infty} E_n(\mathfrak{T}_n)$, <u>where</u> E_n <u>is the linear hull of</u> B_n <u>and</u> $\mathfrak{T}_n$ <u>the topology induced by</u> $\mathfrak{T}$ <u>on</u> E_n .

<u>Remarks.</u> If $\sigma = (E_n)$ in $E(\mathfrak{T})$, where (E_n) is an increasing sequence of subspaces with $E = \bigcup_{n=1}^{\infty} E_n$, then we have $E(\mathfrak{T}_\sigma) = \bigcup_{n=1}^{\infty} E_n(\mathfrak{T}_n)$ ($\mathfrak{T}_n$ is the topology induced by $\mathfrak{T}$). From (7) follows: <u>If</u> $E(\mathfrak{T})$ <u>is countably barrelled, then</u> $E(\mathfrak{T}) = \bigcup_{n=1}^{\infty} E_n(\mathfrak{T}_n)$ <u>for each increasing sequence of subspaces</u> E_n <u>with</u> $E = \bigcup_{n=1}^{\infty} E_n$.

Especially, if $E(\mathfrak{T})$ is an (F)-space, we have $E(\mathfrak{T}) = \bigcup_{n=1}^{\infty} E_n(\mathfrak{T}_n)$, where the $E_n(\mathfrak{T}_n)$ are metrizable. Hence: <u>Each infinite dimensional (F)-space is a strict inductive limit of metrizable spaces.</u>

The following proposition is obvious.

(9) <u>Let</u> $F(\mathfrak{T}_1)$ <u>be a metrizable t.v.s. and let</u> $E(\mathfrak{T}_2)$ <u>be a t.v.s., which has a bornivorous sequence</u> $\sigma = (B_n)$. <u>If</u> $A: F(\mathfrak{T}_1) \longrightarrow E(\mathfrak{T}_2)$ <u>is a continuous linear mapping, then there is a</u> 0-<u>neighbourhood</u> U <u>in</u> $F(\mathfrak{T}_1)$ <u>with</u> $A(U) \subset B_n$ <u>for some</u> n.

In general, in a metrizable t.v.s. there does not exist a bornivorous sequence of bounded sets, since from (9) follows

(10) <u>A metrizable t.v.s.</u> $E(\mathfrak{T})$ <u>admits a bornivorous sequence</u> $\sigma = (B_n)$ <u>of bounded sets, if and only if at least one</u> B_n <u>is a neighbourhood of</u> 0 <u>(i.e. if</u> $E(\mathfrak{T})$ <u>is locally bounded).</u>

At the end of this section we want to consider the completion of spaces with an absorbing sequence. For this we need a more general proposition which is of some interest on its own.

(11) Let H be a dense linear subspace of the t.v.s. $E(\mathfrak{T})$ and let $\hat{\mathfrak{T}}$ be the topology on H induced by $\mathfrak{T}$. Assume that $H(\hat{\mathfrak{T}})$ has an absorbing sequence $\sigma = (B_n)$ with $\hat{\mathfrak{T}} = \hat{\mathfrak{T}}_\sigma$. Then

$$E = \overline{\bigcup_{n=1}^{\infty} B_n}^{\mathfrak{T}} = \bigcup_{n=1}^{\infty} \overline{B_n}^{\mathfrak{T}}$$

and $\bar{\sigma} := (\overline{B_n}^{\mathfrak{T}})$ is an absorbing sequence in $E(\mathfrak{T})$. Furthermore we have $\mathfrak{T} = \mathfrak{T}_{\bar{\sigma}}$.

Proof. If $x \notin \bigcup_{n=1}^{\infty} \overline{B_n}^{\mathfrak{T}}$, then there is for every $n \in \mathbb{N}$ a $\mathfrak{T}$-topological string $\mathfrak{U}^n = (U_j^n)$, such that $x \notin B_n + U_1^n$. Set $W_j := \bigcap_{n=1}^{\infty} (B_{n-j+1} + U_j^n)$. Then the sets $W_j \cap H$ form a $\hat{\mathfrak{T}}_\sigma$-topological string in H by (3),(ii). Hence $(\overline{W_j \cap H}^{\mathfrak{T}})$ is a topological string in $E(\mathfrak{T})$. Since $W_j \supset \overline{W_{j+1} \cap H}^{\mathfrak{T}}$, we obtain that (W_j) is a $\mathfrak{T}$-topological string. For every $n \in \mathbb{N}$ we have $x \notin B_n + W_2$, hence $x \notin \overline{\bigcup_{n=1}^{\infty} B_n}^{\mathfrak{T}}$.

$\mathfrak{T} \subset \mathfrak{T}_{\bar{\sigma}}$ is obvious. Let on the other hand (U_j) be a $\mathfrak{T}_{\bar{\sigma}}$-topological string. By (4) we can assume, that every U_j is $\mathfrak{T}$-closed. $(U_j \cap H)$ is a $\hat{\mathfrak{T}}_\sigma$-topological and hence $\hat{\mathfrak{T}}$-topological string in H. Then $(\overline{U_j \cap H}^{\mathfrak{T}})$ is topological in $E(\mathfrak{T})$, where $(\overline{U_j \cap H}^{\mathfrak{T}}) \subset (U_j)$.

As a corollary of (11) we immediately get a representation of the completion of a space with an absorbing sequence, which contains 5.(11) as a special case.

(12) For a t.v.s. $E(\mathfrak{T})$ let $\tilde{E}(\tilde{\mathfrak{T}})$ be its completion. If $E(\mathfrak{T})$ has an absorbing sequence $\sigma = (B_n)$ with $\mathfrak{T} = \mathfrak{T}_\sigma$, then $\tilde{E} = \bigcup_{n=1}^{\infty} \tilde{B}_n$. If $\tilde{\sigma} := (\tilde{B}_n)$, then $\tilde{\mathfrak{T}} = \tilde{\mathfrak{T}}_{\tilde{\sigma}}$.

A corollary of (12) is

(13) <u>If a t.v.s.</u> $E(\mathfrak{T})$ <u>has an absorbing sequence</u> $\sigma = (B_n)$ <u>of</u> $\mathfrak{T}$<u>-complete</u> B_n , <u>then</u> $E(\mathfrak{T}_\sigma)$ <u>is complete.</u>

We can give another interesting formulation of the property $\tilde{E} = \bigcup_{n=1}^{\infty} \tilde{B}_n$, which in the locally convex case first can be found in G. KÖTHE [1], [4] and D.A. RAIKOV [2] (see N. ADASCH, B. ERNST [3], M. DE WILDE, C. HOUET [1], P. TURPIN [2], M. VALDIVIA [1]).

(14) <u>In a t.v.s.</u> $E(\mathfrak{T})$ <u>with an absorbing sequence</u> $\sigma = (B_n)$ <u>are equivalent</u> ($\mathfrak{U}$ <u>is a base of neighbourhoods of</u> 0 <u>in</u> $E(\mathfrak{T})$):

(i) <u>It is</u> $\widetilde{\bigcup_{n=1}^{\infty} B_n} \subset \bigcup_{n=1}^{\infty} \tilde{B}_n$ <u>(i.e.</u> $\tilde{E} = \bigcup_{n=1}^{\infty} \tilde{B}_n$).

(ii) <u>For each Cauchy filter</u> $\mathfrak{F}$ <u>in</u> $E(\mathfrak{T})$ <u>there is an</u> n_0, <u>such that</u> $\mathfrak{F} + \mathfrak{U} := \{F + U : F \in \mathfrak{F},\ U \in \mathfrak{U}\}$ <u>induces a Cauchy filter on</u> B_{n_0}.

<u>Proof.</u> (i) $\Rightarrow$ (ii): If $\mathfrak{F} \longrightarrow x$ in $\tilde{E}(\tilde{\mathfrak{T}})$, then $x \in \tilde{B}_{n_0}$ for an n_0. Take $F \in \mathfrak{F}$ and a closed neighbourhood U of 0 in $E(\mathfrak{T})$. If V is a balanced 0-neighbourhood with $V + V \subset U$, then $(x + \tilde{V}) \cap B_{n_0} \neq \phi$ and $(x + \tilde{V}) \cap F \neq \phi$, hence $(F + U) \cap B_{n_0} = (F + \tilde{U}) \cap B_{n_0} \neq \phi$.

(ii) $\Rightarrow$ (i): If $x \in \widetilde{\bigcup_{n=1}^{\infty} B_n}$ and $\mathfrak{F}$ is the filter of neighbourhoods of x in $\tilde{E}(\tilde{\mathfrak{T}})$, then $\mathfrak{F}_E := \{F \cap E : F \in \mathfrak{F}\}$ is a Cauchy filter in $E(\mathfrak{T})$. Therefore we have an n_0, such that $\mathfrak{F}_E + \mathfrak{U}$ induces a filter on B_{n_0}. Now, if U and V are neighbourhoods of 0 in $E(\mathfrak{T})$ with $V + V \subset U$, then $(((x + \tilde{V}) \cap E) + V) \cap B_{n_0} \neq \phi$ and $(x + \tilde{U}) \cap B_{n_0} \neq \phi$, that is $x \in \tilde{B}_{n_0}$.

The methods of this section can be used to prove that <u>in a barrelled space a subspace of countable codimension is barrelled.</u>

For this consider in a t.v.s. $E(\mathfrak{T})$ a sequence $(\mathfrak{B}^n)$, where each $\mathfrak{B}^n = (B^n_j)$ is a sequence of balanced sets B^n_j with $B^n_{j+1} + B^n_{j+1} \subset B^n_j$

for all $j \in \mathbb{N}$ and $B_j^n \subset B_j^{n+1}$ for all $n \in \mathbb{N}$. We assume that each $U_j := \bigcup_{n=1}^{\infty} B_j^n$ is absorbing in E, i.e. that (U_j) is a string in E.

Assume $E(\mathcal{T})$ is (countably) barrelled. Then $\overline{\bigcup_{n=1}^{\infty} B_{j+1}^n} \subset \bigcup_{n=1}^{\infty} \overline{B_j^n}$ for all j : If $x \notin \bigcup_{n=1}^{\infty} \overline{B_j^n}$, then there is a sequence $(\mathfrak{U}^n)$ of topological strings $\mathfrak{U}^n = (U_j^n)$ in $E(\mathcal{T})$ with $x \notin \overline{B_j^n + U_1^n}$ for all n . If we set $V_k := \bigcap_{n=1}^{\infty} \overline{B_{j+k}^n + U_{1+k}^n}$, then the string (V_k) is topologically in $E(\mathcal{T})$ by assumption. But $x \notin (\bigcup_{n=1}^{\infty} B_{j+1}^n) + V_1$ and $x \notin \overline{\bigcup_{n=1}^{\infty} B_{j+1}^n}$.

Now consider a barrelled space $E(\mathcal{T})$ and let H be a subspace of countable codimension. For a cobasis (x_n) of H set $H_1 := H \oplus [x_1]$, $H_2 := H \oplus [x_1] \oplus [x_2]$, ...

If $\mathfrak{U} = (U_j)$ is a closed string in H, we can find a closed string $\mathfrak{U}^1 = (U_j^1)$ in H_1 with $\mathfrak{U} = (U_j^1 \cap H)$, a closed string $\mathfrak{U}^2 = (U_j^2)$ in H_2 with $\mathfrak{U}^1 = (U_j^2 \cap H_1)$, ... (see the proof of 6.(10)). $(\bigcup_{n=1}^{\infty} U_j^n)$ is a string in E and $(\overline{\bigcup_{n=1}^{\infty} U_j^n})$ is topological, since $E(\mathcal{T})$ is barrelled. But $\overline{\bigcup_{n=1}^{\infty} U_{j+1}^n} \subset \bigcup_{n=1}^{\infty} \overline{U_j^n}$ as we noted above, and $(\bigcup_{n=1}^{\infty} \overline{U_j^n})$ is topological in $E(\mathcal{T})$. Hence $\mathfrak{U} = (U_j)$ is topological in H, since $U_j = \bigcup_{n=1}^{\infty} \overline{U_j^n} \cap H$.

References: N. ADASCH, B. ERNST [3], M. DE WILDE, C. HOUET [1], [2], W. ROELCKE [1], M. VALDIVIA [1], A. WIWEGER [1].

§ 17 σ-locally topological spaces

In this section we will combine the results of § 15 and § 16. We consider spaces $E(\mathcal{T})$ with an absorbing sequence $\sigma = (B_n)$ of bounded sets B_n. If furthermore $\mathcal{T} = \mathcal{T}_\sigma$ holds, we call the space $E(\mathcal{T})$ σ - l o c a l l y t o p o l o g i c a l . In this case by 16.(6) the sequence $\overline{\sigma} = (\overline{B_n})$ is bornivorous in $E(\mathcal{T})$, and a t.v.s. is σ-locally topological, if and only if it is locally topological and has a bornivorous (= fundamental) sequence of bounded sets. $\mathcal{T} = \mathcal{T}_\sigma$ is the finest linear topology on E, which induces on every balanced bounded set the topology $\mathcal{T}$.

If $E(\mathcal{T})$ is σ-locally topological and σ' is another bornivorous sequence of bounded sets in $E(\mathcal{T})$, then we have $\mathcal{T} = \mathcal{T}_\sigma = \mathcal{T}^{lt} = \mathcal{T}_{\sigma'}$.

We start with a stronger form of 16.(9).

(1) Let $F(\mathcal{T}_2)$ be a metrizable t.v.s. and let $E(\mathcal{T}_1)$ be a σ-locally topological space. If A is a continuous linear mapping from $E(\mathcal{T}_1)$ into $F(\mathcal{T}_2)$ or from $F(\mathcal{T}_2)$ into $E(\mathcal{T}_1)$, then A is strongly bounded *).

Proof. One case is 16.(9). For the other case let (V_n) be a generating string in $F(\mathcal{T}_2)$. Set $U_n := A^{-1}(V_n)$. Then (U_n) is a topological string in $E(\mathcal{T}_1)$. By 16.(3) the string (W_j) with $W_j := \bigcap_{n=1}^{\infty} (B_{n-j+1} + U_{n+j-1})$ is $(\mathcal{T}_1)_\sigma$-topological, hence topological in

*) "strongly bounded" means, there is a neighbourhood of 0, which is mapped onto a bounded set.

$E(\mathfrak{T}_1)$ ((B_n) is a fundamental sequence of bounded sets in $E(\mathfrak{T}_1)$). $A(W_1)$ is absorbed by every V_n .

Now we want to show some permanence properties of σ-locally topological spaces. By 16.(12) we obtain

(2) *If $E(\mathfrak{T})$ is σ-locally topological, then the completion $\tilde{E}(\tilde{\mathfrak{T}})$ is σ-locally topological, and every bounded set in $\tilde{E}(\tilde{\mathfrak{T}})$ is contained in the completion of a bounded set in $E(\mathfrak{T})$.*

From 15.(6) follows

(3) *The topological direct sum $\bigoplus_{i=1}^{\infty} E_i(\mathfrak{T}_i)$ of countably many σ-locally topological spaces $E_i(\mathfrak{T}_i)$ is σ-locally topological.*

(4) *Let $E(\mathfrak{T})$ be σ-locally topological with respect to the fundamental sequence $\sigma = (B_n)$ of bounded sets. Then a quotient space $E/H(\hat{\mathfrak{T}})$ is again σ-locally topological, and if $\hat{\sigma} := (\overline{K_H(B_n)})$, then $\hat{\sigma}$ is a fundamental sequence of bounded sets in $E/H(\hat{\mathfrak{T}})$.*

Proof. It is easy to see, that $\hat{\mathfrak{T}} = \hat{\mathfrak{T}}_{\hat{\sigma}}$. That $\hat{\sigma}$ is bornivorous follows from 16.(6).

(3) and (4) gives

(5) *A countable inductive limit $E(\mathfrak{T}) = \sum_{i=1}^{\infty} A_i(E_i(\mathfrak{T}_i))$ of σ-locally topological spaces $E_i(\mathfrak{T}_i)$ is σ-locally topological.*

For a countable inductive limit of σ-locally topological spaces we can prove a rather strong proposition on bounded sets (see 5.(9)).

(6) If $E(\mathfrak{T}) = \sum_{i=1}^{\infty} A_i(E_i(\mathfrak{T}_i))$ is an inductive limit of the σ-locally topological spaces $E_i(\mathfrak{T}_i)$, then there exist for every bounded set B in $E(\mathfrak{T})$ a $k \in \mathbb{N}$ and bounded sets $B_i \subset E_i(\mathfrak{T}_i)$, $1 \leq i \leq k$, such that $B \subset \overline{\sum_{i=1}^{k} A_i(B_i)}$.

The proof is an easy consequence of (3) and (4) (see 4.(5)).

We now give an "outer characterization" of σ-locally topological spaces, i.e. a characterization by some properties of certain spaces of linear mappings. First we show

(7) The t.v.s. $E(\mathfrak{T})$ has a fundamental sequence $\sigma = (B_n)$ of bounded sets, if and only if for every (F)-space $F(\mathfrak{T}')$ the space $L_b(E,F)$ is metrizable.

Proof. If $L_b(E,F)$ is metrizable for every (F)-space $F(\mathfrak{T}')$, we construct a certain (F)-space similar as in the proof of 7.(2) . Let $\{(U_j^\gamma) : \gamma \in \Gamma\}$ be a set of strings, such that $\{U_1^\gamma : \gamma \in \Gamma\}$ is a base of neighbourhoods of 0 in $E(\mathfrak{T})$. Set $E_\gamma := E$ for all $\gamma \in \Gamma$, $F' := \bigoplus_{\gamma \in \Gamma} E_\gamma$, $V_j := \bigoplus_{\gamma \in \Gamma} U_{j+2}^\gamma$. $\mathfrak{W} = (V_j)$ is a string in F'. Let $F(\mathfrak{T}')$ be the completion of $F'_{\mathfrak{W}}$. Then the string $(\widetilde{K(V_j)})$ generates $\mathfrak{T}'$, where K is the canonical mapping from F' into F.

Now it suffices to show, that for every bounded set B in $E(\mathfrak{T})$ and all $i,j \in \mathbb{N}$ the sets

$$[B,\widetilde{K(V_i)}]^{-1}(\widetilde{K(V_j)}) := \bigcap \{A^{-1}(\widetilde{K(V_j)}) : A \in [B,\widetilde{K(V_i)}]\}$$

are bounded in $E(\mathfrak{T})$. To show this let I_γ, $\gamma \in \Gamma$, be the canonical embedding of E into $\bigoplus_{\gamma \in \Gamma} E_\gamma$ and set $\hat{I}_\gamma := K \circ I_\gamma$. Then $\hat{I}_\gamma \in L(E,F)$ and there is a $\lambda_\gamma > 0$, such that $\lambda_\gamma \hat{I}_\gamma \in [B,\widetilde{K(V_i)}]$. Then

$$[B,\widetilde{K(V_i)}]^{-1}(\widetilde{K(V_j)}) \subset (\lambda_\gamma \hat{I}_\gamma)^{-1}(\widetilde{K(V_j)})$$
$$\subset \lambda_\gamma^{-1} I_\gamma^{-1}(V_j + V_j + V_j)$$
$$\subset \lambda_\gamma^{-1} U_j^t .$$

Since every bounded B is contained in $[B,\widetilde{K(V_j)}]^{-1}(\widetilde{K(V_j)})$, and since $L_b(E,F)$ is metrizable by hypothesis, we can find a fundamental sequence of bounded sets in $E(\mathfrak{T})$.

Now we can give the new characterization of σ-locally topological spaces.

(8) A t.v.s. $E(\mathfrak{T})$ is σ-locally topological, if and only if for every (F)-space $F(\mathfrak{T}')$ the following conditions hold:

a) $L_b(E,F)$ is metrizable, and b) $L_b(E,F)$ is complete.

Proof. By (7) condition a) is equivalent with the existence of a fundamental sequence $\sigma = (B_n)$ of bounded sets in $E(\mathfrak{T})$, and the completeness of $L_b(E,F)$ follows by 15.(10), if $E(\mathfrak{T})$ is σ-locally topological.

To prove the converse (we have to show $\mathfrak{T} = \mathfrak{T}_\sigma$), we will again construct a suitable (F)-space. Let $\mathfrak{W} = (V_j)$ be a $\mathfrak{T}_\sigma$-topological string. It suffices to show that $\mathfrak{W}$ is also $\mathfrak{T}$-topological. By 16.(3) we can assume, that the V_j are of the form $V_j = \bigcap_{n=1}^{\infty} (B_{n-j+1} + U_j^n)$, where $(U_j^n)_{j\in\mathbb{N}}$ is for every $n \in \mathbb{N}$ a $\mathfrak{T}$-topological string.

Set $E_i := E$ for all $i \in \mathbb{N}$, $F' := \{(x_i) \in \prod_{i=1}^{\infty} E_i : x_i = x$ for some $x \in E$ and nearly all $i \in \mathbb{N}\}$ and $S_j := F' \cap \prod_{n=1}^{\infty} (B_{n-j+1} + U_j^n)$. $\mathfrak{S} = (S_j)$ is a string in F'. Let $F(\mathfrak{T}')$ be the completion of $F'_{\mathfrak{S}}$.

Let I_m be the linear mapping from E into F' defined by $I_m(x) := (x,\ldots,\underset{m^{th}}{x},0,0,\ldots)$. Then $I_m(V_j) \subset S_j$. The mappings $\hat{I}_m := K \circ I_m$

are continuous mappings from $E(\mathfrak{T})$ into $F(\mathfrak{T}')$ (K is the canonical mapping from F' into $\widetilde{F'_\mathfrak{S}} = F$). Let $I: E \longrightarrow F'$ be defined by $I(x) := (x, x, \ldots)$ and set $\hat{I} := K \circ I$.

For a given bounded set B_n in $E(\mathfrak{T})$ and a neighbourhood $\widetilde{K(S_j)}$ in $F(\mathfrak{T}')$ we have $(\hat{I} - \hat{I}_{n+j-1})(B_n) \subset K(S_j)$, and $\hat{I} = \hat{I}_{n+j-1} + (\hat{I} - \hat{I}_{n+j-1})$ where $\hat{I}_{n+j-1} \in L(E,F)$. Since $L_b(E,F)$ is complete by hypothesis, $\hat{I}$ is continuous by 12.(6) . That means $\hat{I}^{-1}((\widetilde{K(S_j)}))$ is a topological string in $E(\mathfrak{T})$. On the other hand we have $\hat{I}^{-1}(\widetilde{K(S_{j+2})}) \subset I^{-1}(S_j) = V_j$. Hence $\mathfrak{W} = (V_j)$ is topological in $E(\mathfrak{T})$.

Another formulation of (8) gives a quite interesting proposition, namely

(9) <u>The class of σ-locally topological spaces is the maximal class of t.v.s. $E(\mathfrak{T})$, such that for every (F)-space $F(\mathfrak{T}')$ the space $L_b(E,F)$ is again an (F)-space.</u>

At the end of this section we want to consider once more locally topological spaces. We want to show that every locally topological space can be represented as an inductive limit of "elementary" locally topological spaces, that means as an inductive limit of σ-locally topological spaces.

If B is balanced and bounded in the t.v.s. $E(\mathfrak{T})$, consider the set $B_n := \overline{\sum_{i=1}^{2^{n-1}} B}$ and denote by E_B the linear hull of the B_n , i.e. $E_B := \bigcup_{n=1}^{\infty} B_n$. The sequence $\sigma_B = (B_n)$ is absorbing in E_B . Denote by $\hat{\mathfrak{T}}$ the topology induced by $\mathfrak{T}$ on E_B , then we can form the topology $\hat{\mathfrak{T}}_{\sigma_B}$. Since every B_n is $\mathfrak{T}$-bounded, it is also $\hat{\mathfrak{T}}_{\sigma_B}$-bounded (see 15.(3)), and we obtain with 16.(6)

(10) $\sigma_B = (B_n)$ <u>forms a fundamental sequence of bounded sets in</u> <u>$E_B(\widehat{\mathcal{T}}_{\sigma_B})$. $E_B(\widehat{\mathcal{T}}_{\sigma_B})$ is σ-locally topological.</u>

(11) <u>For an arbitrary t.v.s. $E(\mathcal{T})$</u> <u>holds</u> $E(\mathcal{T}^{lt}) = \sum_{B\in\mathfrak{B}} E_B(\widehat{\mathcal{T}}_{\sigma_B})$, <u>where $\mathfrak{B}$ is a fundamental system of bounded sets in $E(\mathcal{T})$.</u> <u>If $E(\mathcal{T})$ is locally topological, we have</u> $E(\mathcal{T}) = \sum_{B\in\mathfrak{B}} E_B(\widehat{\mathcal{T}}_{\sigma_B})$.

In a similar way we can construct "elementary" bornological spaces. For a bounded balanced set B in a t.v.s. $E(\mathcal{T})$ let $\mathcal{T}_B$ be the finest linear topology on E_B , for which all B_n are bounded. A string $\mathfrak{U} = (U_j)$ is $\mathcal{T}_B$-topological, if every U_j absorbs the B_n . $E_B(\mathcal{T}_B)$ is bornological. (10) implies $(\widehat{\mathcal{T}}_{\sigma_B})^\beta = \mathcal{T}_B$ and therefore (B_n) is a fundamental sequence of bounded sets in $E_B(\mathcal{T}_B)$. Now we have

(12) <u>It is</u> $E(\mathcal{T}^\beta) = \sum_{B\in\mathfrak{B}} E_B(\mathcal{T}_B)$ <u>for each t.v.s. $E(\mathcal{T})$. If $E(\mathcal{T})$ is bornological, then $E(\mathcal{T})$ is the inductive limit of these "elementary" bornological spaces</u> $E_B(\mathcal{T}_B)$.

If B is absolutely convex, then $E_B(\mathcal{T}_B)$ is the well known normed space with B as unit ball.

References: N. ADASCH, B. ERNST [4].

§ 18 (DF)-spaces and spaces with a fundamental sequence of compact sets

A. GROTHENDIECK introduced the class of (DF)-spaces using for their definition properties of the strong dual of a locally convex (F)-space. In our terminology we can give GROTHENDIECK's definition in the following way: A locally convex space $E(\mathfrak{T})$ is a (DF)-space, if it is countably quasibarrelled in $\mathcal{C}$ and has a fundamental sequence of bounded sets.

It is obvious how to extend this notion to the class $\mathcal{L}$ of all topological vector spaces: A t.v.s. $E(\mathfrak{T})$ is a (DF) - space (in $\mathcal{L}$), if $E(\mathfrak{T})$ is countably quasibarrelled (in $\mathcal{L}$) and has a fundamental sequence of bounded sets.

By 16.(7) every (DF)-space is σ-locally topological, but the converse fails to be true as a counter example shows (19.6.).

Another characterization of (DF)-spaces can be obtained by combining 13.(7) and 17.(7). The rest of the usual theory of (DF)-spaces is in a more general form contained in § 15, § 16 and § 17.

In this section we consider spaces $E(\mathfrak{T})$ with a fundamental sequence of compact sets, that means spaces with an absorbing sequence $\sigma = (B_n)$ of compact sets B_n, such that every compact subset of $E(\mathfrak{T})$ is contained in some B_n.

First we note, that such a fundamental sequence $\sigma = (B_n)$ of compact sets B_n is always bornivorous in $E(\mathfrak{T})$: assume there is a bounded balanced set B contained in no B_n, then there are $x_n \in \frac{1}{n}B$ with $x_n \notin B_n$. The sequence (x_n) is contained in no B_n, but on the other hand $\{(x_n), 0\}$ is a compact subset.

To prove some interesting properties of σ-locally topological spaces with a fundamental sequence of compact sets, we need the following fundamental theorem.

(1) Let $E(\mathfrak{T})$ be a t.v.s. with an absorbing sequence $\sigma = (B_n)$ of closed B_n, such that $\mathfrak{T} = \mathfrak{T}_\sigma$. If M is a subset of $E(\mathfrak{T})$, such that $M \cap B_n$ is compact for all $n \in \mathbb{N}$, then M is closed.

Proof. If $x \notin M$, we have to show $x \notin \overline{M}$. We assume $x \in B_1$. There is a closed balanced 0-neighbourhood U^1, such that $(x+U^1) \cap (M \cap B_2) = \emptyset$, hence *)

$$(x + (U^1 \cap B_1)) \cap M = \emptyset ,$$

and $(x + (U^1 \cap B_1)) \cap (M \cap B_3) = \emptyset$. Since $x + (U^1 \cap B_1)$ is closed and $M \cap B_3$ compact, we find a closed balanced 0-neighbourhood U^2 with $(x + (U^1 \cap B_1) + U^2 + U^2) \cap (M \cap B_3) = \emptyset$, especially

$$(x + (U^1 \cap B_1) + (U^2 \cap B_2)) \cap M = \emptyset .$$

Since $\overline{(x + (U^1 \cap B_1) + (U^2 \cap B_2))} \cap (M \cap B_3) = \emptyset$ and $\overline{(x + (U^1 \cap B_1) + (U^2 \cap B_2))} \subset B_1 + \overline{B_1 + B_2} \subset B_3$, even follows $\overline{(x + (U^1 \cap B_1) + (U_2 \cap B_2))} \cap M = \emptyset$ and $\overline{(x + (U^1 \cap B_1) + (U^2 \cap B_2))} \cap (M \cap B_4) = \emptyset$.

Repeating this procedure we obtain a sequence (U^n) of 0-neighbourhoods, such that

$$(x + \sum_{n=1}^{\infty} (B_n \cap U^n)) \cap M = \emptyset .$$

With $\mathfrak{T} = \mathfrak{T}_\sigma$ and 16.(3) follows $x \notin \overline{M}$.

Using 16.(12) one sees immediately that M in (1) is a complete subset of $E(\mathfrak{T})$.

Furthermore, if $\mathfrak{T}'$ is a topology on M which coincides on all sets $M \cap B_n$ with $\mathfrak{T}$, then (1) implies $\mathfrak{T}' \subset \mathfrak{T}|_M$. Especially we have

---

*) We recall that $B_n + B_n \subset B_{n+1}$. Furthermore we will set again $B_n := \{0\}$ for $n \leq 0$.

(2) If $E(\mathfrak{T})$ is a σ-locally topological space with a fundamental sequence $\sigma = (B_n)$ of compact B_n, then the closed subsets M of $E(\mathfrak{T})$ are those, for which all $M \cap B_n$ are closed. $\mathfrak{T}$ is the finest of all topologies on E, which coincide on every B_n with $\mathfrak{T}$.

Another interesting consequence of (1) is the following: For a σ-locally topological space with a fundamental sequence of compact sets each closed subspace has the same properties.

σ-locally topological spaces with a fundamental sequence of compact sets are interesting with respect to the closed graph theorem and the open mapping theorem, since we can show

(3) If $E(\mathfrak{T}_0)$ is σ-locally topological with a fundamental sequence $\sigma = (B_n)$ of compact sets, then $E(\mathfrak{T}_0)$ is B-complete.

Proof. Since every quotient space of $E(\mathfrak{T}_0)$ is again σ-locally topological with a fundamental sequence of compact sets (see 17.(4)), it suffices to show that $E(\mathfrak{T}_0)$ is B_r-complete. Let $\mathfrak{T}$ be a Hausdorff linear topology on E with $\mathfrak{T} \subset \mathfrak{T}_0$ and $\overline{\mathfrak{T}_0}^{\mathfrak{T}} \subset \mathfrak{T}$. Then the B_n are also $\mathfrak{T}$-compact, that means $\mathfrak{T}_\sigma = (\mathfrak{T}_0)_\sigma = \mathfrak{T}_0$. Since $\mathfrak{T}_\sigma$ has by 16.(4) a base of neighbourhoods of 0 of $\mathfrak{T}$-closed sets, we have $\mathfrak{T}_0 = \mathfrak{T}_\sigma \subset \overline{\mathfrak{T}_0}^{\mathfrak{T}} \subset \mathfrak{T}$.

We even have the following general theorem:

(4) A t.v.s. $E(\mathfrak{T}_0)$ with an absorbing sequence $\sigma = (B_n)$ of compact B_n is an s-space.

Proof. The assumption on $E(\mathfrak{T}_0)$ remains true for each quotient space, so it suffices to prove, that $E(\mathfrak{T}_0)$ is an infra-s-space. For this let $\mathfrak{T}$ be a Hausdorff linear topology on E with $\mathfrak{T} \subset \mathfrak{T}_0$. Then $\mathfrak{T}$ and $\mathfrak{T}_0$ coincide on each B_n and we have $\mathfrak{T}_\sigma = (\mathfrak{T}_0)_\sigma$. But $((\mathfrak{T}_0)_\sigma)^t = \mathfrak{T}_0^{\,t}$ and $(\mathfrak{T}_\sigma)^t = \mathfrak{T}^t$ (for each $\mathfrak{T}$ and σ we have $(\mathfrak{T}_\sigma)^t = \mathfrak{T}^t$, since $\mathfrak{T} \subset \mathfrak{T}_\sigma \subset \mathfrak{T}^t$ by 16.(4)), therefore $\mathfrak{T}^t = \mathfrak{T}_0^{\,t}$.

We want to prove now some results for (DF)-spaces, in which all bounded sets are relatively compact.

(5) Let B be a precompact subset of the countably quasibarrelled (countably barrelled) space $E(\mathfrak{T})$. Then B is $\mathfrak{T}^{b^*}$-precompact ($\mathfrak{T}^b$-precompact).

Proof. We prove only one assertion. The proof of the other one is analogous.

Assume B were not $\mathfrak{T}^{b^*}$-precompact. Then there exists a closed bornivorous string $\mathfrak{U} = (U_j)$ in $E(\mathfrak{T})$, such that for all finite subsets $\{x_1,\dots,x_n\} \subset B$ we have $B \not\subset \bigcup_{i=1}^{n} (x_i + U_1)$. That means, we can find a sequence (y_n) in B with $y_n \notin \{y_1,\dots,y_{n-1}\} + U_1$ for all $n > 1$. Since $0 \notin -y_n + \{y_1,\dots,y_{n-1}\} + U_1$ and $-y_n + \{y_1,\dots,y_{n-1}\} + U_1$ is closed, there are topological strings $\mathfrak{W}_n = (V_j^n)$ in $E(\mathfrak{T})$, such that

$$V_1^n \cap (-y_n + \{y_1,\dots,y_{n-1}\} + U_1) = \emptyset .$$

Hence

$$\overline{(V_2^n + U_1)} \cap (-y_n + \{y_1,\dots,y_{n-1}\}) = \emptyset .$$

Now the string (W_j) with $W_j := \bigcap_{n=1}^{\infty} \overline{V_{j+1}^n + U_j}$ is a topological string in $E(\mathfrak{T})$, and for all $n > 1$

$$W_1 \cap (-y_n + \{y_1,\dots,y_{n-1}\}) = \emptyset .$$

The sequence (y_n) contains a Cauchy subsequence (y_{n_i}). Hence for certain i we have $-y_{n_i}+y_{n_{i-1}} \in W_1$ and

$$-y_{n_i}+y_{n_{i-1}} \in (-y_{n_i}+\{y_1,\dots,y_{n_i-1}\}) \cap W_1 = \emptyset .$$

If the set B in (5) is compact, then B is also $\mathcal{T}^{b*}$-compact ($\mathcal{T}^{b}$-compact) by G. KÖTHE [4], § 18, 4.(4) and we see at once with (2)

(6) A (DF)-space with a fundamental sequence of compact sets is quasibarrelled.

This is not true for σ-locally topological spaces with a fundamental sequence of compact sets (see 19.6.).

If $E(\mathcal{T})$ is a t.v.s. with a fundamental sequence $\mathfrak{S} = (B_n)$ of bounded sets, then the strings $\mathfrak{V} = (V_j)$ with $V_j := \sum_{n=1}^{\infty} \lambda_n B_{n-j+1}$, where the (λ_n) are positive sequences, form a base of 0-neighbourhoods for the associated bornological topology $\mathcal{T}^{\beta}$. With this remark we can prove

(7) Let $E(\mathcal{T}_0)$ be a (DF)-space with the fundamental sequence $\mathfrak{S} = (B_n)$ of bounded sets. If the B_n are compact for some Hausdorff linear topology $\mathcal{T}$ on E with $\mathcal{T} \subset \mathcal{T}_0$, then the topologies $\mathcal{T}_0^{\beta}$ and $\mathcal{T}_0^{b*}$ coincide.

Proof. For a (V_j) with $V_j = \sum_{n=1}^{\infty} \lambda_n B_{n-j+1}$ we prove $\overline{V_2}^{\mathcal{T}_0} \subset V_1$. If $x_0 \notin V_1$, then $x_0 \notin V_2 + V_2$, and therefore $(x_0 + \sum_{n=1}^{m} \lambda_n B_{n-1}) \cap (\sum_{n=1}^{m} \lambda_n B_{n-1}) = \emptyset$ for all $m \in \mathbb{N}$. Since both sets are $\mathcal{T}$-compact, there are $\mathcal{T}$-topological strings $\mathfrak{W}_m = (W_j^m)$, such that we have $\overline{(x_0 + \sum_{n=1}^{m} \lambda_n B_{n-1} + W_1^m}^{\mathcal{T}_0}) \cap (\sum_{n=1}^{m} \lambda_n B_{n-1}) = \emptyset$. Now for every m the strings (T_j^m) with $T_j^m := \overline{\sum_{n=1}^{m} \lambda_n B_{n-j} + W_j^m}^{\mathcal{T}_0}$ and the string (T_j) with $T_j := \bigcap_{m=1}^{\infty} T_j^m$ are $\mathcal{T}_0$-topological. But $(x_0 + T_1) \cap V_2 = \emptyset$, that means $x_0 \notin \overline{V_2}^{\mathcal{T}_0}$.

A consequence of (7) and (6) is

(8) A (DF)-space with a fundamental sequence of compact sets is bornological.

In certain cases we can improve (6) in another direction. Since a space with a fundamental sequence of compact sets is sequentially complete, we have with (6) and 14.(6)

(9) A boundedly summing (DF)-space with a fundamental sequence of compact sets is barrelled.

We shall now show that most vector spaces with a fundamental sequence of compact sets are "nearly" (DF)-spaces.

(10) Let $E(\mathcal{T})$ be a t.v.s. with a fundamental sequence $\sigma = (B_n)$ of compact B_n. Then $E(\mathcal{T}^{b*})$ is a (DF)-space with σ as fundamental sequence of bounded sets.
Furthermore we have $(\mathcal{T}^{b*})^{b*} = \mathcal{T}^{qt} = \mathcal{T}^{\beta}$, i.e. the quasibarrelled and the bornological topology associated to $\mathcal{T}$ coincide.

Proof. $E(\mathcal{T}^{b*})$ will be a (DF)-space, if $\mathcal{T}^{b*}$ is countably quasibarrelled.

Let $\mathfrak{U}_n = (U_j^n)$ be a sequence of $\mathcal{T}^{b*}$-closed $\mathcal{T}^{b*}$-topological strings, such that each $U_j := \bigcap_{n=1}^{\infty} U_j^n$ is bornivorous in $E(\mathcal{T}^{b*})$. Then there are positive numbers λ_j^n, such that $\lambda_j^n B_{n-j} \subset U_{n+j+1}$ and $\lambda_j^n \geq \lambda_{j+1}^n$ for all $j,n \in \mathbb{N}$. We obtain $\sum_{i=1}^{n} \lambda_j^i B_{i-j} \subset U_{j+1}$ for all $n \in \mathbb{N}$. For each (U_j^n) we can find a $\mathcal{T}$-bornivorous $\mathcal{T}$-closed string (W_j^n), such that $W_j^n \subset U_{j+1}^n$ for all $n,j \in \mathbb{N}$. Define

$$V_j^n := W_j^n + \sum_{i=1}^{n} \lambda_j^i B_{i-j} \quad .$$

Then each (V_j^n) is a closed string in $E(\mathcal{T})$.

Each V_j with $V_j := \bigcap_{n=1}^{\infty} V_j^n$ is bornivorous in $E(\mathcal{T})$: if B is bounded, then $B \subset B_{n_0-j}$ for an $n_0 \in \mathbb{N}$, and

$$V_j = \bigcap_{n=1}^{n_0-1} (W_j^n + \sum_{i=1}^{n} \lambda_j^i B_{i-j}) \cap \bigcap_{n=n_0}^{\infty} (W_j^n + \sum_{i=1}^{n} \lambda_j^i B_{i-j}) \quad ,$$

$$\lambda_j^{n_0} B_{n_0-j} \subset \bigcap_{n=n_0}^{\infty} (W_j^n + \sum_{i=1}^{n} \lambda_j^i B_{i-j}) \quad .$$

Hence (V_j) is a closed bornivorous string in $E(\mathcal{T})$ and a topological string in $E(\mathcal{T}^{b*})$. On the other hand we have

$$\begin{aligned} V_j &\subset W_j^n + \sum_{i=1}^{n} \lambda_j^i B_{i-j} \\ &\subset U_{j+1}^n + U_{j+1} \subset U_{j+1}^n + U_{j+1}^n \subset U_j^n \end{aligned}$$

for all $n \in \mathbb{N}$, therefore $V_j \subset U_j$. (U_j) is a topological string in $E(\mathcal{T}^{b*})$.

The last remark of the theorem follows with (7) applied to the space $E(\mathcal{T}^{b*})$.

In (10) the B_n need not further be $\mathcal{T}^{b*}$-compact, for instance, consider a reflexive Banach space.

As an easy corollary of (10) we obtain in certain cases an analogous result for the strong topology:

(11) If $E(\mathfrak{T})$ is boundedly summing and has a fundamental sequence $\sigma = (B_n)$ of compact B_n, then $E(\mathfrak{T}^b)$ is a (DF)-space with σ as fundamental sequence of bounded sets. We have $\mathfrak{T}^\beta = (\mathfrak{T}^b)^b = \mathfrak{T}^t$.

Proof. $E(\mathfrak{T})$ is sequentially complete and we have $\mathfrak{T}^{b*} = \mathfrak{T}^b$ by 14.(6). Therefore $E(\mathfrak{T}^b)$ is a (DF)-space by (10) with $\mathfrak{T}^\beta = (\mathfrak{T}^{b*})^{b*} = (\mathfrak{T}^b)^{b*}$. $(\mathfrak{T}^b)^{b*} = \mathfrak{T}^\beta$ is boundedly summing and quasibarrelled, and $(\mathfrak{T}^b)^{b*}$ is sequentially complete by G. KÖTHE [4], § 18, 4.(4), hence $(\mathfrak{T}^b)^{b*}$ is barrelled by 14.(6). This implies $(\mathfrak{T}^b)^{b*} = (\mathfrak{T}^b)^b = \mathfrak{T}^t$.

At the end of this section we prove a homomorphism theorem, which in the locally convex case was shown by G. KÖTHE [2] for (DF)-spaces with fundamental sequences of compact sets. First we need

(12) If $E(\mathfrak{T})$ is a sequentially complete and boundedly summing t.v.s. with absorbing sequences $\sigma = (B_n)$ and $\tau = (C_n)$ of bounded sets B_n and C_n, then $\mathfrak{T}_\sigma = \mathfrak{T}_\tau$.

Proof. We assume that the B_n and C_n are closed sets (see 16.(5)). By 16.(4) $\mathfrak{T}_\sigma$ is generated by a set of $\mathfrak{T}$-closed strings. But each $\mathfrak{T}$-closed string absorbs the bounded sets in $E(\mathfrak{T})$ (see 14.(6)), and therefore the C_n are bounded in $\mathfrak{T}_\sigma$. $\sigma = (B_n)$ is bornivorous in $E(\mathfrak{T}_\sigma)$ by 16.(6), hence we have $C_n \subset B_k$ for all n and certain k. From this follows that $\mathfrak{T}_\sigma$ induces on C_n the same topology as $\mathfrak{T}$, i.e. $\mathfrak{T}_\sigma \subset \mathfrak{T}_\tau$ and $\mathfrak{T}_\sigma = \mathfrak{T}_\tau$ by symmetry.

Now we can prove G. KÖTHE's homomorphism theorem.

(13) Let $E(\mathfrak{T}_1)$ and $F(\mathfrak{T}_2)$ be σ-locally topological and boundedly summing spaces with fundamental sequences of compact sets. A continuous linear mapping $A: E(\mathfrak{T}_1) \longrightarrow F(\mathfrak{T}_2)$ is a topological homomorphism, if and only if $A(E)$ is closed in $F(\mathfrak{T}_2)$.

Proof. If A is a homomorphism, then $E(\mathfrak{T}_1)/N(A)$ is isomorphic to $A(E)(\widehat{\mathfrak{T}}_2)$. But $E(\mathfrak{T}_1)/N(A)$ is complete by 17.(4) and 16.(13), hence $A(E)(\widehat{\mathfrak{T}}_2)$ is complete and $A(E)$ is closed in $F(\mathfrak{T}_2)$.

Conversely let A(E) be closed. We can assume, that A is injective and hence E a subspace of F (see 17.(4)). If $\widehat{\mathfrak{T}}_2 := \mathfrak{T}_2|_E$, then $E(\widehat{\mathfrak{T}}_2)$ is again boundedly summing and σ-locally topological with a fundamental sequence σ_2 of compact sets (see the remark after (2)). Especially $E(\widehat{\mathfrak{T}}_2)$ is complete.

Since A is continuous we have $\mathfrak{T}_1 \supset \widehat{\mathfrak{T}}_2$. If σ_1 is a compact fundamental sequence in $E(\mathfrak{T}_1)$, then σ_1 is an absorbing sequence of compact sets in $E(\widehat{\mathfrak{T}}_2)$, hence $(\mathfrak{T}_1)_{\sigma_1} = (\widehat{\mathfrak{T}}_2)_{\sigma_1}$. With (12) follows $(\widehat{\mathfrak{T}}_2)_{\sigma_1} = (\widehat{\mathfrak{T}}_2)_{\sigma_2}$ and $\mathfrak{T}_1 = (\mathfrak{T}_1)_{\sigma_1} = (\widehat{\mathfrak{T}}_2)_{\sigma_1} = (\widehat{\mathfrak{T}}_2)_{\sigma_2} = \widehat{\mathfrak{T}}_2$.

As an easy application of the last sections we give some results of R. WAGNER [1] (and D.A. RAIKOV [1], J. SEBASTIAO E SILVA [1] in the locally convex case).

We consider a sequence of t.v.s. $E_k(\mathfrak{T}_k)$, such that $E_k \subset E_{k+1}$ and the embeddings $E_k(\mathfrak{T}_k) \longrightarrow E_{k+1}(\mathfrak{T}_{k+1})$ are compact*) for all $k \in \mathbb{N}$. Let $\mathfrak{T}$ be the finest linear topology on $E := \bigcup_{k=1}^{\infty} E_k$, such that all embeddings of the $E_k(\mathfrak{T}_k)$ are continuous.

---

*) That means there is a neighbourhood of 0 in $E_k(\mathfrak{T}_k)$, which is relatively compact in $E_{k+1}(\mathfrak{T}_{k+1})$. This implies the continuity of the embeddings.

$\mathfrak{F}$ <u>is Hausdorff:</u> To see this it is enough to construct for $x \in E_1$ with $x \neq 0$ a $\mathfrak{F}$-topological string $\mathfrak{U} = (U_n)$ with $x \notin U_1$. First there is a closed 0-neighbourhood V_2 in $E_2(\mathfrak{F}_2)$ with $x \notin V_2$. Further there is a topological string (V_n^1) in $E_1(\mathfrak{F}_1)$, such that $\overline{V_n^1}^{\mathfrak{F}_2}$ is compact in $E_2(\mathfrak{F}_2)$ and $\overline{V_1^1}^{\mathfrak{F}_2} \subset V_2$. Set $U_n^1 := \overline{V_n^1}^{\mathfrak{F}_2}$. Then $x \notin U_1^1$ and, since U_1^1 is compact in $E_3(\mathfrak{F}_3)$, $x \notin U_1^1 + V_3$ for a closed neighbourhood V_3 of 0 in $E_3(\mathfrak{F}_3)$. Choose a topological string (V_n^2) in $E_2(\mathfrak{F}_2)$, such that $\overline{V_n^2}^{\mathfrak{F}_3}$ is compact in $E_3(\mathfrak{F}_3)$ and $\overline{V_1^2}^{\mathfrak{F}_3} \subset V_3$. Then $x \notin U_1^1 + U_1^2$, if $U_n^2 := \overline{V_n^2}^{\mathfrak{F}_3}$, and $U_1^1 + U_1^2$ is compact in $E_4(\mathfrak{F}_4)$. In this way we get a sequence of strings $\mathfrak{U}^k = (U_n^k)$. $\mathfrak{U} = (U_n)$ with $U_n := \sum_{k=1}^{\infty} U_n^k$ is by 4.(3) a topological string in $E(\mathfrak{F})$, but $x \notin U_1$.

Such a c o m p a c t i n d u c t i v e l i m i t $E(\mathfrak{F})$ has very strong properties.

(i) $E(\mathfrak{F})$ <u>has an absorbing sequence</u> $\sigma = (B_n)$ <u>of compact</u> B_n :
Set $B_1 := \overline{U_1}^{\mathfrak{F}_2}$, $B_n := \overline{B_{n-1} \cup U_n}^{\mathfrak{F}_{n+1}} + \overline{B_{n-1} \cup U_n}^{\mathfrak{F}_{n+1}}$ for $n \geq 2$, where U_n is a balanced neighbourhood of 0 in $E_n(\mathfrak{F}_n)$, which is relatively compact in $E_{n+1}(\mathfrak{F}_{n+1})$.

(ii) <u>For each</u> $B_n \in \sigma$ <u>there is a sequence</u> (λ_j) <u>with</u> $\lambda_j > 0$, <u>such that</u> $\sum_{j=1}^{\infty} \lambda_j B_n$ <u>is relatively compact in</u> $E(\mathfrak{F})$: Choose a topological string (U_j) in $E_{n+1}(\mathfrak{F}_{n+1})$ with $U_1 \subset B_{n+1}$. Since B_n is compact in $E_{n+1}(\mathfrak{F}_{n+1})$, we have $\lambda_j B_n \subset U_{j+1}$ with $\lambda_j > 0$ and therefore $\sum_{j=1}^{\infty} \lambda_j B_n \subset \sum_{j=1}^{\infty} U_{j+1} \subset U_1 \subset B_{n+1}$, which is compact in $E(\mathfrak{F})$.

(iii) $E(\mathfrak{F})$ <u>is a</u> σ<u>-locally topological space with</u> $\sigma = (B_n)$ <u>as</u>

a fundamental sequence of compact sets. $E(\mathcal{T})$ *is boundedly summing. It is B-complete:* The B_n are $\mathcal{T}_\sigma$-compact, hence the embeddings $E_k(\mathcal{T}_k) \longrightarrow E(\mathcal{T}_\sigma)$ are continuous. This implies $\mathcal{T} = \mathcal{T}_\sigma$. That $\sigma = (B_n)$ is a fundamental sequence follows with 16.(6). With (ii) follows that $E(\mathcal{T})$ is boundedly summing, and with (3) that it is B-complete.

(iv) $E(\mathcal{T})$ *is bornological, barrelled and a (DF)-space:* Each bornivorous string in $E(\mathcal{T})$ induces topological strings on $E_k(\mathcal{T}_k)$, hence it is topological in $E(\mathcal{T})$ by 4.(1), i.e. $E(\mathcal{T})$ is bornological. Furthermore it is complete and boundedly summing, hence barrelled by 14.(6).

Finally, immediate consequences of (iii) and (2) are

(v) $B \subset E(\mathcal{T})$ *is bounded, if and only if it is bounded in an* $E_k(\mathcal{T}_k)$. $M \subset E(\mathcal{T})$ *is closed, if and only if all* $M \cap E_k$ *are closed in* $E_k(\mathcal{T}_k)$. *If* E_k *is closed in* $E_{k+1}(\mathcal{T}_{k+1})$ *for all* k *, then the* E_k *are closed in* $E(\mathcal{T})$. $\mathcal{T}$ *is the finest topology on* E, *for which all embeddings of the* $E_k(\mathcal{T}_k)$ *are continuous.*

References: N. ADASCH, B. ERNST [2], B. ERNST [1], B. ERNST, R. WAGNER [1], S.O. IYAHEN [2], J.P. LIGAUD [1].

§ 19 Some examples and counter examples

In this section we want to show for some of the introduced classes of topological vector spaces, that they are really different from each other, furthermore we are interested in the relation of these classes to the corresponding classes of locally convex spaces.

Since the notions of barrelled spaces, bornological spaces, ... , s-spaces, infra-s-spaces, ... , as we used them up to here, are based on the category $\mathcal{L}$ of topological vector spaces, we emphasize this by speaking of $\mathcal{L}$-barrelled, $\mathcal{L}$-bornological spaces, ... , s-spaces in $\mathcal{L}$, infra-s-spaces in $\mathcal{L}$, ...

The corresponding notions in the category $\mathcal{C}$ of locally convex spaces will now be $\mathcal{C}$-barrelled, $\mathcal{C}$-bornological, ... , s-space in $\mathcal{C}$, infra-s-space in $\mathcal{C}$, ... For their definitions as far as they are not given here see G. KÖTHE [4].

1. It is trivial that each $\mathcal{L}$-barrelled locally convex space is also $\mathcal{C}$-barrelled. The converse is false as can be shown by simple examples. For instance, if $\mathfrak{T}^c \neq \mathfrak{T}^f$ for E, then $E(\mathfrak{T}^c)$ is a $\mathcal{C}$-barrelled space, which is not $\mathcal{L}$-barrelled (the strings $\mathfrak{U}^p$ in § 1 are $\mathfrak{T}^c$-closed, but they are not $\mathfrak{T}^c$-topological for $0 < p < 1$).

If $\mathfrak{T}$ is a linear topology, we will denote by $\mathfrak{T}^{\infty}$ its *associated locally convex topology*. $\mathfrak{T}^{\infty}$ is generated by the absolutely convex 0-neighbourhoods of $\mathfrak{T}$ (and therefore $\mathfrak{T}^{\infty}$ is the finest locally convex topology coarser than $\mathfrak{T}$). If $\mathfrak{T}^{\infty}$ is Hausdorff (this is not necessarily the case), then $E(\mathfrak{T}^{\infty})$ is called the *locally convex space associated to* $E(\mathfrak{T})$.

It is easy to see, that the associated locally convex space $E(\mathfrak{T}^{\infty})$ of an $\mathcal{L}$-barrelled space $E(\mathfrak{T})$ is $\mathcal{C}$-barrelled.

A simple example of a normed $\mathcal{C}$-barrelled space, which is not $\mathcal{L}$-barrelled can be given in the following way (see W. ROBERTSON [1]): For $0 < p < 1$ we denote by $(\ell^p)_1$ the sequence space ℓ^p endowed with the norm topology of ℓ^1. If $(\ell^p)^{oo}$ is the associated locally convex space of the (F)-space ℓ^p, then we have $(\ell^p)^{oo} = (\ell^p)_1$. Therefore $(\ell^p)_1$ is $\mathcal{C}$-barrelled, but it is not $\mathcal{L}$-barrelled, since the " ℓ^p-balls" are closed in $(\ell^p)_1$ without being O-neighbourhoods.

With the usual topologies the spaces $\mathcal{E}$ (= the space of all infinitely differentiable functions) and $\mathcal{S}$ (= the space of all rapidly decreasing functions in $\mathcal{E}$) of test functions are (F)-spaces, hence they are barrelled in $\mathcal{L}$. The space $\mathcal{D}$ (= the space of all functions of $\mathcal{E}$ with compact support) is barrelled in $\mathcal{L}$ as a countable inductive limit (in $\mathcal{L}$, see 4.(6)) of certain (F)-spaces $\mathcal{D}(K)$, K a compact subset of $\mathbb{R}^n$.

The strong duals $\mathcal{E}'$, $\mathcal{S}'$, $\mathcal{D}(K)'$ are (DF)-spaces (in $\mathcal{C}$ or in $\mathcal{L}$, see 6. in this section) with a fundamental sequence of compact sets. The following proposition (or 18.(9)) implies that they are barrelled in $\mathcal{L}$.

(1) <u>If $E(\mathcal{T})$ is a locally convex space with a fundamental sequence (B_n) of bounded sets, where the B_n are Banach disks, then $E(\mathcal{T})$ is barrelled in $\mathcal{L}$, if and only if it is barrelled in $\mathcal{C}$.</u>

<u>Proof.</u> Assume $E(\mathcal{T})$ is barrelled in $\mathcal{C}$. If $\mathfrak{U} = (U_n)$ is a closed string in $E(\mathcal{T})$, then $\mathfrak{U}$ absorbs each B_n by 6.(6). With $\lambda_n > 0$, such that $\lambda_n B_n \subset U_n$, we have

$$\sum_{n=m+1}^{\infty} \lambda_n B_n \subset \sum_{n=m+1}^{\infty} U_n \subset U_m \quad \text{and} \quad \overline{\sum_{n=m+1}^{\infty} \lambda_n B_n} \subset U_m .$$

$\overline{\sum_{n=m+1}^{\infty} \lambda_n B_n}$ is a barrel in $E(\mathcal{T})$ and therefore a neighbourhood of O. Hence $\mathfrak{U}$ is a topological string.

The space $\mathcal{D}'$ of distributions is a complemented subspace of a product of spaces $\mathcal{D}(K)'$ (see D. KEIM [2]), hence it is also barrelled in $\mathcal{L}$.

2. We have similar relations as in 1. for bornological spaces. $\mathcal{L}$-bornological locally convex spaces are always $\mathcal{C}$-bornological, but the converse is not true in general, as can be shown again by the first example in 1. . Nevertheless, if we make further assumptions, we can prove

(2) Let $E(\mathcal{T})$ be a locally convex space with a fundamental sequence (B_n) of bounded subsets. If $E(\mathcal{T})$ is $\mathcal{C}$-bornological ($\mathcal{C}$-quasibarrelled), then $E(\mathcal{T})$ is also $\mathcal{L}$-bornological ($\mathcal{L}$-quasibarrelled).

We prove (2) for the quasibarrelled case: For a bornivorous closed string $\mathcal{U} = (U_n)$ in $E(\mathcal{T})$ we have $\sum_{n=m+1}^{\infty} \lambda_n B_n \subset U_m$ as in the proof of (1). Since $E(\mathcal{T})$ is $\mathcal{C}$-quasibarrelled, $\sum_{n=m+1}^{\infty} \lambda_n B_n$ is a neighbourhood of 0 (we may assume, that the B_n are absolutely convex), and $\mathcal{U}$ is a topological string in $E(\mathcal{T})$.

Later on we will prove the same proposition for d-quasibarrelled spaces, a class greater than the class of quasibarrelled spaces. For barrelled spaces (2) is wrong in this general form (see the second example in 1.).

Again $\mathcal{E}$, $\mathcal{S}$, $\mathcal{D}(K)$, $\mathcal{D}$ are bornological in $\mathcal{L}$, and the same is true for $\mathcal{E}'$, $\mathcal{S}'$, $\mathcal{D}(K)'$ by (2). $\mathcal{D}'$ is bornological in $\mathcal{L}$ as a complemented subspace of a countable product of $\mathcal{L}$-bornological spaces (see 11.(7)).

3. Now we give examples, which distinguish the class of bornological t.v.s. from the class of barrelled t.v.s..

There are t.v.s. which are $\mathcal{L}$-bornological, but not $\mathcal{L}$-barrelled, as is shown by the first example in § 6 .

But there are also $\mathcal{L}$-barrelled spaces, which are not $\mathcal{L}$-bornological: For $d > \aleph_0$ let Π be a product of d $\mathcal{L}$-barrelled spaces and Π_0 the subspace of those elements, which have at most countably many components different from 0 . As 6.(9) one can prove that Π_0 is $\mathcal{L}$-barrelled, and therefore each subspace Π_1 with $\Pi_0 \subset \Pi_1 \subset \Pi$ is $\mathcal{L}$-barrelled by 6.(12). For $x_0 \in \Pi \setminus \Pi_0$ we now prove that $\Pi_1 :=$ $\Pi_0 + [x_0]$ is not $\mathcal{L}$-bornological. Since Π_0 is sequentially closed in Π, it follows (see Y. KOMURA [1], proof of prop. 5.2), that each bounded subset of Π_1 is contained in a set $B_0 + [x_0]_\lambda$, B_0 bounded in Π_0. If Π_1 would be $\mathcal{L}$-bornological, then the bornivorous absolutely convex set $\Pi_0 + [x_0]_1$ would be a 0-neighbourhood in Π_1, and $\bigcap_{\varepsilon>0} \varepsilon(\Pi_0 + [x_0]_1) = \Pi_0$ had to be closed in Π_1. This cannot be, since Π_0 is dense in Π.

Of course, this construction gives also $\mathcal{C}$-barrelled locally convex spaces, which are not $\mathcal{C}$-bornological.

Furthermore these examples show, that the classes of $\mathcal{L}$-barrelled and $\mathcal{L}$-bornological spaces are proper subclasses of the $\mathcal{L}$-quasibarrelled spaces.

4. Here we stay in the category $\mathcal{L}$. We give P. TURPIN's [2] example of a t.v.s. which is bornological (hence quasibarrelled) and complete, but not barrelled (see also L. WAELBROECK [1]). This example is very important showing some aspects of the theory of t.v.s. unexpected from the theory of locally convex spaces (compare § 14).

Let φ be the space of all finite sequences and $e_j \in \varphi$ the jth unit vector. If $B := \{e_j : j \in \mathbb{N}\}$, then denote by $\mathcal{T}$ the finest linear topology on φ for which B is bounded. $\varphi(\mathcal{T})$ has the desired properties.

Of course $\varphi(\mathcal{T})$ is a bornological space. $\varphi(\mathcal{T})$ is not barrelled: Consider the continuous linear mappings A_k in $\varphi(\mathcal{T})$ with $A_k(x) :=$ $\sum_{j=1}^{k} j\, \xi_j e_j$, if $x = (\xi_j) \in \varphi$. The sequence (A_k) converges pointwise to a linear mapping A . If $\varphi(\mathcal{T})$ would be barrelled, A would be continuous by 7.(6). This cannot be since $A(B) = \{je_j : j \in \mathbb{N}\}$ is unbounded in $\varphi(\mathcal{T})$.

$\varphi(\mathcal{T})$ is complete: The sets $B_n := \{(\xi_j) : |\{j : \xi_j \neq 0\}| \leq 2^n$ and $|\xi_j| \leq 2^n\}$ are $\mathcal{T}$-bounded, since they are subsets of certain linear combinations of B . The B_n are also $\mathcal{T}^\infty$-bounded ($\mathcal{T}^\infty$ is the topology generated by the sup-norm on φ), hence $(\mathcal{T}^\infty)_\sigma$-bounded by 15.(3), where $\sigma = (B_n)$. This implies $(\mathcal{T}^\infty)_\sigma \subset \mathcal{T}$. $(\mathcal{T}^\infty)_\sigma = \mathcal{T}$ follows, since $\mathcal{T}$ induces on B_n the same 0-neighbourhoods as $\mathcal{T}^\infty$. The B_n are $\mathcal{T}^\infty$-complete and therefore $\varphi((\mathcal{T}^\infty)_\sigma) = \varphi(\mathcal{T})$ is complete by 16.(13).

5. Now we will turn to those spaces, which are important for the closed graph theorem and the open mapping theorem.

For the locally convex B_r- resp. B-complete spaces (here called B_r- resp. B-complete in $\mathcal{C}$) and infra-s- resp. s-spaces see V. PTAK [1], Y. KOMURA [1], N. ADASCH [1].

A locally convex space $E(\mathcal{T}_0)$ is called *infra-s-space in* $\mathcal{C}$, if for each weaker locally convex Hausdorff topology $\mathcal{T}$ on E the associated $\mathcal{C}$-barrelled topologies $\mathcal{T}^{ct}$ and $\mathcal{T}_0^{ct}$ coincide.

A locally convex space $E(\mathcal{T}_0)$ is called *B_r-complete in* $\mathcal{C}$, if or each weaker locally convex Hausdorff topology $\mathcal{T}$ on E the relation $\overline{\mathcal{T}_0}^{\mathcal{T}} \subset \mathcal{T}$ implies $\mathcal{T}_0 = \mathcal{T}$ (see 10.(3)).

If every quotient space of $E(\mathcal{T}_0)$ is an infra-s-space in $\mathcal{C}$ (a B_r-complete space in $\mathcal{C}$), then the locally convex space $E(\mathcal{T}_0)$ is called *s-space in* $\mathcal{C}$ (*B-complete in* $\mathcal{C}$).

One conclusion of the following proposition is clear at once:

(3) <u>A locally convex space $E(\mathcal{T}_0)$ is B- (resp. B_r-) complete in $\mathcal{C}$, if and only if it is B- (resp. B_r-) complete in $\mathcal{L}$.</u>

We prove that the B_r-completeness in $\mathcal{C}$ of $E(\mathcal{T}_0)$ implies the B_r-completeness in $\mathcal{L}$. For this choose a linear Hausdorff topology $\mathcal{T}$ on E with $\mathcal{T} \subset \mathcal{T}_0$ and $\overline{\mathcal{T}_0}^{\mathcal{T}} \subset \mathcal{T}$, i.e. $\overline{\mathcal{T}_0}^{\mathcal{T}} = \mathcal{T}$. $\overline{\mathcal{T}_0}^{\mathcal{T}}$ and therefore $\mathcal{T}$ is locally convex, and since $E(\mathcal{T}_0)$ is B_r-complete in $\mathcal{C}$ follows $\mathcal{T} = \mathcal{T}_0$.

Only one half of a proposition corresponding to (3) is true for s- resp. infra-s-spaces.

Consider the second example in 1. , the space $(\ell^p)_1$ with $0 < p < 1$. $(\ell^p)_1$ is $\mathcal{C}$-barrelled. It is not complete, hence not B- or B_r-complete in $\mathcal{C}$. Therefore it cannot be an s- or infra-s-space in $\mathcal{C}$. But $(\ell^p)_1$ is an s-space in $\mathcal{L}$, because its topology is coarser than the topology of the s-space ℓ^p.

Therefore there are s- resp. infra-s-spaces in $\mathcal{L}$, which are neither s- nor infra-s-spaces in $\mathcal{C}$. On the other hand we have (V. EBERHARDT [1])

(4) <u>Each locally convex s- (resp. infra-s-) space in $\mathcal{C}$ is an s- (resp. infra-s-) space in $\mathcal{L}$.</u>

<u>Proof.</u> We have to consider the case of infra-s-spaces. If $E(\mathcal{T}_0)$ is an infra-s-space in $\mathcal{C}$, and if $\mathcal{T}$ is a coarser Hausdorff linear topology on E , then we have

$$\mathcal{T} \subset \overline{\mathcal{T}_0}^{\mathcal{T}} \subset \mathcal{T}_0 \quad \text{and} \quad \overline{\mathcal{T}_0}^{\mathcal{T}} \subset \mathcal{T}^t .$$

Since $\overline{\mathcal{T}_0}^{\mathcal{T}}$ is a Hausdorff locally convex topology on E and $E(\mathcal{T}_0)$ is an infra-s-space in $\mathcal{C}$, follows

$$\mathcal{T}_0 \subset \mathcal{T}_0^{ct} = (\overline{\mathcal{T}_0}^{\mathcal{T}})^{ct} \subset (\overline{\mathcal{T}_0}^{\mathcal{T}})^t = \mathcal{T}^t ,$$

hence $\mathcal{T}_0^t = \mathcal{T}^t$, i.e. $E(\mathcal{T}_0)$ is an infra-s-space in $\mathcal{L}$.

6. The notion of locally topological spaces can also be introduced within the category of locally convex spaces.

A locally convex space $E(\mathfrak{T})$ is said to be **locally topological in $\mathfrak{C}$** (or **$\mathfrak{C}$-locally topological**), if each locally continuous linear mapping from $E(\mathfrak{T})$ into a locally convex space $F(\mathfrak{T}')$ is continuous (see K. NOUREDDINE [1]).

One can see, that we have properties of these spaces and results analogous to those of § 15 and § 17. For instance, a locally convex space is locally topological in $\mathfrak{C}$, if and only if every absolutely convex set is a 0-neighbourhood, which intersects each bounded balanced subset in a 0-neighbourhood of the induced topology.

Of course, each $\mathfrak{L}$-locally topological locally convex space is $\mathfrak{C}$-locally topological, and the converse is not true without further assumptions. So it follows from 16.(3)

(5) Let $E(\mathfrak{T})$ be a locally convex space with a fundamental sequence of bounded sets. If $E(\mathfrak{T})$ is $\mathfrak{C}$-locally topological, then $E(\mathfrak{T})$ is $\mathfrak{L}$-locally topological.

This result suggests the conjecture, that a similar proposition is true for (DF)-spaces.

In this context (only) we write "(DF)-spaces in $\mathfrak{C}$" for the locally convex (DF)-spaces introduced by A. GROTHENDIECK to distinguish the "(DF)-spaces in $\mathfrak{L}$" of § 18.

Each locally convex (DF)-space in $\mathfrak{L}$ is a (DF)-space in $\mathfrak{C}$. The converse is given by a more general proposition (N. ADASCH [6]).

(6) A locally convex space $E(\mathfrak{T})$ with a fundamental sequence (B_n) of bounded sets is d-quasibarrelled in $\mathfrak{C}$, if and only if it is d-quasibarrelled in $\mathfrak{L}$.

Proof. Let $E(\mathfrak{T})$ be d-quasibarrelled in $\mathfrak{C}$. Let $\mathfrak{U} = (U_n)$ be a bornivorous string in $E(\mathcal{T})$, which is intersection of d closed topological strings $\mathfrak{U}^\alpha = (U_n^\alpha)$, i.e. $U_n = \bigcap_\alpha U_n^\alpha$. We find $\lambda_n > 0$ with $\lambda_n B_n \subset U_n$ for all $n \in \mathbb{N}$. Therefore $\lambda_n B_n \subset U_n^\alpha$ for all α and $n \in \mathbb{N}$. If V_n^α is an absolutely convex neighbourhood of 0 in $E(\mathcal{T})$ with $V_n^\alpha \subset U_n^\alpha$ for all α and $n \in \mathbb{N}$, we have

$$W_n^\alpha := \lambda_{n+1} B_{n+1} + V_{n+1}^\alpha \subset U_{n+1}^\alpha + U_{n+1}^\alpha \subset U_n^\alpha$$

and

$$\sum_{n=m+1}^{\infty} \lambda_{n+1} B_{n+1} \subset \sum_{n=m+1}^{\infty} W_n^\alpha \subset \sum_{n=m+1}^{\infty} U_n^\alpha \subset U_m^\alpha .$$

From this follows for all α and $m \in \mathbb{N}$

$$\sum_{n=m+1}^{\infty} \lambda_{n+1} B_{n+1} \subset \overline{\sum_{n=m+1}^{\infty} W_n^\alpha} \subset U_m^\alpha ,$$

hence

$$\sum_{n=m+1}^{\infty} \lambda_{n+1} B_{n+1} \subset V_m := \bigcap_\alpha \left(\overline{\sum_{n=m+1}^{\infty} W_n^\alpha} \right)$$

$$\subset \bigcap_\alpha U_m^\alpha = U_m .$$

Therefore each V_m is bornivorous and intersection of d closed absolutely convex 0-neighbourhoods (we assume the B_n absolutely convex). Since $E(\mathcal{T})$ is d-quasibarrelled in $\mathcal{C}$, the V_m and the U_m are neighbourhoods of 0 in $E(\mathcal{T})$.

The following example (B. ERNST [1]) distinguishes the locally topological t.v.s. from the (DF)-spaces and the bornological spaces.

For this let $E(\mathcal{T})$ be a (not finite dimensional) reflexive Banach space, K its $\mathcal{T}_s(E(\mathcal{T})')$-compact unit ball. Set $\sigma := (nK : n \in \mathbb{N})$. We consider $E(\mathcal{T}_s(E(\mathcal{T})'))$ and the topology $(\mathcal{T}_s(E(\mathcal{T})'))_\sigma$ on E (see § 16). For this locally convex space we write $E(\mathcal{T}_0)$. $E(\mathcal{T}_0)$ is a

σ-locally topological space with a fundamental sequence of compact sets. From 18.(2) follows, that $\mathcal{T}_0$ is the finest topology on E, which on each nK , $n \in \mathbb{N}$, coincides with the weak topology $\mathcal{T}_s(E(\mathcal{T})')$. If we apply Banach-Dieudonne's theorem (G. KÖTHE [4], § 21, 10.(1)) to the space $E(\mathcal{T})'(\mathcal{T}_k(E))$ ($\mathcal{T}_k(E)$ is the Mackey topology on E'), then $\mathcal{T}_0$ coincides with the topology $\mathcal{T}_c = \mathcal{T}_c(E(\mathcal{T})'(\mathcal{T}_k(E)))$, the topology of uniform convergence on all $\mathcal{T}_k(E)$-precompact subsets of the dual space $E(\mathcal{T})'$.

Now we see at once, that $E(\mathcal{T}_0)$ cannot be quasibarrelled, hence not bornological, since otherwise we had $\mathcal{T}_k(E(\mathcal{T})') = \mathcal{T}_0 = \mathcal{T}_c$.

Furthermore $E(\mathcal{T}_0)$ is no (DF)-space, for this would imply by proposition 18.(8), that $E(\mathcal{T}_0)$ is bornological.

References

N. Adasch: [1] Tonnelierte Räume und zwei Sätze von Banach. Math.Ann. 186, 209-214 (1970).

[2] Topologische Produkte gewisser topologischer Vektorräume. Math.Ann. 189, 280-284 (1970).

[3] Über die Vollständigkeit von $L_{\sigma}(E,F)$. Math.Ann. 191, 290-292 (1971).

[4] Der Graphensatz in topologischen Vektorräumen. Math. Z. 119, 131-142 (1971).

[5] Vollständigkeit und der Graphensatz. J.reine angew. Math. 249, 217-220 (1971).

[6] Lokalkonvexe Räume mit einer Fundamentalfolge beschränkter Teilmengen. Math.Ann. 199, 257-261 (1972).

[7] Über lokaltopologische Vektorräume. Proc.Symposium Funct.Analysis, Istanbul (Silivri) 1973.

N. Adasch, B. Ernst: [1] Teilräume gewisser topologischer Vektorräume. Collectanea Math. XXIV, 27-39 (1973).

[2] Ultra-(DF)-Räume mit relativ kompakten beschränkten Teilmengen. Math.Ann. 206, 79-87 (1973).

[3] Lokaltopologische Vektorräume. Collectanea Math. XXV, 255-274 (1974).

[4] Lokaltopologische Vektorräume II. Collectanea Math. XXVI, 13-18 (1975).

S. Banach: [1] Théorie des operations linéaires. Warszawa 1932.

N. Bourbaki: [1] Éléments de mathématique, Livre V: Espaces vectoriels topologiques. Paris: Hermann 1966.

M. De Wilde: [1] Vector topologies and linear maps on products of topological vector spaces. Math.Ann. 196, 117-128 (1972).

M. De Wilde, C. Houet: [1] On increasing sequences of absolutely convex sets in locally convex spaces. Math.Ann. 192, 257-261 (1971).

[2] Sur les propriétés de tonnelage des espaces vectoriels topologiques. Bull.Soc.Roy.Sci.Liège 40, 555-560 (1971).

V. Eberhardt: [1] Einige Vererbbarkeitseigenschaften von B- und B_r-vollständigen Räumen. Math.Ann. 215, 1-11 (1975).

B. Ernst: [1] Ultra-(DF)-Räume. J.reine angew.Math. 258, 87-102 (1973).

B. Ernst, R. Wagner: [1] Räume mit einer absorbierenden Folge kompakter Mengen. J.reine angew.Math. 278/279, 398-407 (1975).

B. Gramsch: [1] Ein Zerlegungssatz für Resolventen elliptischer Operatoren. Math.Z. 133, 219-242 (1973).

A. Grothendieck: [1] Sur les espaces (F) et (DF). Summa Brasil. Math. 3, 57-123 (1954).

[2] Espaces vectoriels topologiques. São Paulo 1954.

S.O. Iyahen: [1] On certain classes of linear topological spaces. Proc.London Math.Soc. 18, 285-307 (1968).

[2] Linear topological spaces with fundamental sequences of compact sets. Math.Ann. 200, 179-183 (1973).

S. Kakutani: [1] Über die Metrisation der topologischen Gruppen. Proc.Imp.Acad.Tokyo 12, 82-84 (1936).

D. Keim: [1] Die Ordnungstopologie und ordnungstonnelierte Topologien auf Vektorverbänden. Collectanea Math. XXII, 117-140 (1971).

[2] Induktive und projektive Limiten mit Zerlegung der Einheit. Manuscripta math. 10, 191-195 (1973).

J. Kelley, I. Namioka e.a.: [1] Linear topological spaces. New York - London - Toronto: D. van Nostrand 1963.

J. Köhn: [1] Induktive Limiten nicht lokalkonvexer Räume. Math. Ann. 181, 269-278 (1969).

Y. Komura: [1] On linear topological spaces. Kumamoto J.Sci. Series A, 5, No.3, 148-157 (1962).

G. Köthe: [1] Über die Vollständigkeit einer Klasse lokalkonvexer Räume. Math.Z. 52, 627-630 (1950).

[2] Homomorphismen von (F)-Räumen. Math.Z. 84, 219-221 (1964).

[3] General linear transformations of locally convex spaces. Math.Ann. 159, 309-328 (1965).

[4] Topologische lineare Räume I. Berlin - Heidelberg - New York: Springer 1966.

J.P. Ligaud: [1] Espaces DF non nécessairement localement convexes. C.R.Acad.Sci.Paris 275, 283-285 (1972).

M. Mahowald: [1] Barrelled spaces and the closed graph theorem. J.London Math.Soc. 36, 108-110 (1961).

K. Noureddine: [1] Nouvelles classes d'espaces localement convexes. C.R.Acad.Sci.Paris 276, 1209-1212 (1973).

K. Noureddine, J. Schmets: [1] Espaces associés à un espace localement convexe et espaces de fonctions continues. Bull. Soc.Roy.Sc.Liège 42, 116-124 (1973).

A. Pietsch: [1] Nukleare lokalkonvexe Räume. Berlin: Akademie-Verlag 1969.

H. Pfister: [1] Über die Stetigkeit von linearen Abbildungen auf Produkten von topologischen linearen Räumen. To appear.

V. Ptak: [1] Completeness and the open mapping theorem. Bull. Soc.Math.France 86, 41-74 (1958).

D.A. Raikov: [1] Vollstetige Spektren lokalkonvexer Räume (russisch). Trudy Moskov.Mat.Obsc. 7, 413-438 (1958).

[2] Vollständigkeitskriterien für lokalkonvexe Räume (russisch). Uspeki Mat.Nauk. 14.1, 223-229 (1959).

W. Robertson: [1] Completions of topological vector spaces. Proc.London Math.Soc. 8, 242-257 (1958).

[2] On the closed graph theorem and spaces with webbs. Proc.London Math.Soc. 24, 692-738 (1972).

A.P. and W. Robertson: [1] On the closed graph theorem. Proc. Glasgow Math.Assoc. 3, 9-12 (1956).

[2] Topological vector spaces. Cambridge: University Press 1964.

W. Roelcke: [1] On the finest locally convex topology agreeing with a given topology on a sequence of absolutely convex sets. Math.Ann. 198, 57-80 (1972).

S. Rolewicz: [1] Metric linear spaces. Warszawa 1972.

W. Ruess, R. Wagner: [1] Über beschränkte Mengen in induktiven Limiten topologischer Vektorräume. Manuscripta math. 19, 365-374 (1976).

J. Sebastiao e Silva: [1] Su certe classi di spazi localmente convessi importanti per le applicazioni. Rend.Mat.Roma V, ser. 14, 388-410 (1955).

H.H. Schaefer: [1] Topological vector spaces. Berlin - Heidelberg - New York: Springer 1971.

S. Tomašek: [1] M-barrelled spaces. Comment.Math.Univ.Carolinae 11, 185-204 (1970).

[2] M-bornological spaces. Comment.Math.Univ.Carolinae 11, 235-248 (1970).

P. Turpin: [1] Généralisation d'un théorème de S.Mazur et W.Orlicz. C.R.Acad.Sci.Paris 273, 457-460 (1971).

[2] Convexités dans les espaces vectoriels topologiques généraux. Thèse, Orsay 1974, published in: Dissertationes Mathematicae CXXXI, Warszawa 1976.

M. Valdivia: [1] Absolutely convex sets in barrelled spaces. Ann.Inst.Fourier 21, 3-13 (1971).

L. Waelbroeck: [1] Topological vector spaces and algebras. Lecture Notes in Math. 230. Berlin - Heidelberg - New York: Springer 1971.

R. Wagner: [1] Topologisch lineare induktive Limiten mit abzählbarem kompakten Spektrum. J.reine angew.Math. 261, 209-215 (1973).

A. Wiweger: [1] Linear spaces with mixed topology. Studia Math. 20, 47-68 (1961).

Subject index

Vol. 460: O. Loos, Jordan Pairs. XVI, 218 pages. 1975.

Vol. 461: Computational Mechanics. Proceedings 1974. Edited by J. T. Oden. VII, 328 pages. 1975.

Vol. 462: P. Gérardin, Construction de Séries Discrètes p-adiques. »Sur les séries discrètes non ramifiées des groupes réductifs déployés p-adiques«. III, 180 pages. 1975.

Vol. 463: H.-H. Kuo, Gaussian Measures in Banach Spaces. VI, 224 pages. 1975.

Vol. 464: C. Rockland, Hypoellipticity and Eigenvalue Asymptotics. III, 171 pages. 1975.

Vol. 465: Séminaire de Probabilités IX. Proceedings 1973/74. Edité par P. A. Meyer. IV, 589 pages. 1975.

Vol. 466: Non-Commutative Harmonic Analysis. Proceedings 1974. Edited by J. Carmona, J. Dixmier and M. Vergne. VI, 231 pages. 1975.

Vol. 467: M. R. Essén, The Cos $\pi\lambda$ Theorem. With a paper by Christer Borell. VII, 112 pages. 1975.

Vol. 468: Dynamical Systems – Warwick 1974. Proceedings 1973/74. Edited by A. Manning. X, 405 pages. 1975.

Vol. 469: E. Binz, Continuous Convergence on C(X). IX, 140 pages. 1975.

Vol. 470: R. Bowen, Equilibrium States and the Ergodic Theory of Anosov Diffeomorphisms. III, 108 pages. 1975.

Vol. 471: R. S. Hamilton, Harmonic Maps of Manifolds with Boundary. III, 168 pages. 1975.

Vol. 472: Probability-Winter School. Proceedings 1975. Edited by Z. Ciesielski, K. Urbanik, and W. A. Woyczyński. VI, 283 pages. 1975.

Vol. 473: D. Burghelea, R. Lashof, and M. Rothenberg, Groups of Automorphisms of Manifolds. (with an appendix by E. Pedersen) VII, 156 pages. 1975.

Vol. 474: Séminaire Pierre Lelong (Analyse) Année 1973/74. Edité par P. Lelong. VI, 182 pages. 1975.

Vol. 475: Répartition Modulo 1. Actes du Colloque de Marseille-Luminy, 4 au 7 Juin 1974. Edité par G. Rauzy. V, 258 pages. 1975. 1975.

Vol. 476: Modular Functions of One Variable IV. Proceedings 1972. Edited by B. J. Birch and W. Kuyk. V, 151 pages. 1975.

Vol. 477: Optimization and Optimal Control. Proceedings 1974. Edited by R. Bulirsch, W. Oettli, and J. Stoer. VII, 294 pages. 1975.

Vol. 478: G. Schober, Univalent Functions – Selected Topics. V, 200 pages. 1975.

Vol. 479: S. D. Fisher and J. W. Jerome, Minimum Norm Extremals in Function Spaces. With Applications to Classical and Modern Analysis. VIII, 209 pages. 1975.

Vol. 480: X. M. Fernique, J. P. Conze et J. Gani, Ecole d'Eté de Probabilités de Saint-Flour IV–1974. Edité par P.-L. Hennequin. XI, 293 pages. 1975.

Vol. 481: M. de Guzmán, Differentiation of Integrals in R^n. XII, 226 pages. 1975.

Vol. 482: Fonctions de Plusieurs Variables Complexes II. Séminaire François Norguet 1974–1975. IX, 367 pages. 1975.

Vol. 483: R. D. M. Accola, Riemann Surfaces, Theta Functions, and Abelian Automorphisms Groups. III, 105 pages. 1975.

Vol. 484: Differential Topology and Geometry. Proceedings 1974. Edited by G. P. Joubert, R. P. Moussu, and R. H. Roussarie. IX, 287 pages. 1975.

Vol. 485: J. Diestel, Geometry of Banach Spaces – Selected Topics. XI, 282 pages. 1975.

Vol. 486: S. Stratila and D. Voiculescu, Representations of AF-Algebras and of the Group U (∞). IX, 169 pages. 1975.

Vol. 487: H. M. Reimann und T. Rychener, Funktionen beschränkter mittlerer Oszillation. VI, 141 Seiten. 1975.

Vol. 488: Representations of Algebras, Ottawa 1974. Proceedings 1974. Edited by V. Dlab and P. Gabriel. XII, 378 pages. 1975.

Vol. 489: J. Bair and R. Fourneau, Etude Géométrique des Espaces Vectoriels. Une Introduction. VII, 185 pages. 1975.

Vol. 490: The Geometry of Metric and Linear Spaces. Proceedings 1974. Edited by L. M. Kelly. X, 244 pages. 1975.

Vol. 491: K. A. Broughan, Invariants for Real-Generated Uniform Topological and Algebraic Categories. X, 197 pages. 1975.

Vol. 492: Infinitary Logic: In Memoriam Carol Karp. Edited by D. W. Kueker. VI, 206 pages. 1975.

Vol. 493: F. W. Kamber and P. Tondeur, Foliated Bundles and Characteristic Classes. XIII, 208 pages. 1975.

Vol. 494: A Cornea and G. Licea. Order and Potential Resolvent Families of Kernels. IV, 154 pages. 1975.

Vol. 495: A. Kerber, Representations of Permutation Groups II. V, 175 pages. 1975.

Vol. 496: L. H. Hodgkin and V. P. Snaith, Topics in K-Theory. Two Independent Contributions. III, 294 pages. 1975.

Vol. 497: Analyse Harmonique sur les Groupes de Lie. Proceedings 1973–75. Edité par P. Eymard et al. VI, 710 pages. 1975.

Vol. 498: Model Theory and Algebra. A Memorial Tribute to Abraham Robinson. Edited by D. H. Saracino and V. B. Weispfenning. X, 463 pages. 1975.

Vol. 499: Logic Conference, Kiel 1974. Proceedings. Edited by G. H. Müller, A. Oberschelp, and K. Potthoff. V, 651 pages 1975.

Vol. 500: Proof Theory Symposion, Kiel 1974. Proceedings. Edited by J. Diller and G. H. Müller. VIII, 383 pages. 1975.

Vol. 501: Spline Functions, Karlsruhe 1975. Proceedings. Edited by K. Böhmer, G. Meinardus, and W. Schempp. VI, 421 pages. 1976.

Vol. 502: János Galambos, Representations of Real Numbers by Infinite Series. VI, 146 pages. 1976.

Vol. 503: Applications of Methods of Functional Analysis to Problems in Mechanics. Proceedings 1975. Edited by P. Germain and B. Nayroles. XIX, 531 pages. 1976.

Vol. 504: S. Lang and H. F. Trotter, Frobenius Distributions in GL_2-Extensions. III, 274 pages. 1976.

Vol. 505: Advances in Complex Function Theory. Proceedings 1973/74. Edited by W. E. Kirwan and L. Zalcman. VIII, 203 pages. 1976.

Vol. 506: Numerical Analysis, Dundee 1975. Proceedings. Edited by G. A. Watson. X, 201 pages. 1976.

Vol. 507: M. C. Reed, Abstract Non-Linear Wave Equations. VI, 128 pages. 1976.

Vol. 508: E. Seneta, Regularly Varying Functions. V, 112 pages. 1976.

Vol. 509: D. E. Blair, Contact Manifolds in Riemannian Geometry. VI, 146 pages. 1976.

Vol. 510: V. Poènaru, Singularités C^∞ en Présence de Symétrie. V, 174 pages. 1976.

Vol. 511: Séminaire de Probabilités X. Proceedings 1974/75. Edité par P. A. Meyer. VI, 593 pages. 1976.

Vol. 512: Spaces of Analytic Functions, Kristiansand, Norway 1975. Proceedings. Edited by O. B. Bekken, B. K. Øksendal, and A. Stray. VIII, 204 pages. 1976.

Vol. 513: R. B. Warfield, Jr. Nilpotent Groups. VIII, 115 pages. 1976.

Vol. 514: Séminaire Bourbaki vol. 1974/75. Exposés 453 – 470. IV, 276 pages. 1976.

Vol. 515: Bäcklund Transformations. Nashville, Tennessee 1974. Proceedings. Edited by R. M. Miura. VIII, 295 pages. 1976.

Vol. 516: M. L. Silverstein, Boundary Theory for Symmetric Markov Processes. XVI, 314 pages. 1976.

Vol. 517: S. Glasner, Proximal Flows. VIII, 153 pages. 1976.

Vol. 518: Séminaire de Théorie du Potentiel, Proceedings Paris 1972–1974. Edité par F. Hirsch et G. Mokobodzki. VI, 275 pages. 1976.

Vol. 519: J. Schmets, Espaces de Fonctions Continues. XII, 150 pages. 1976.

Vol. 520: R. H. Farrell, Techniques of Multivariate Calculation. X, 337 pages. 1976.

Vol. 521: G. Cherlin, Model Theoretic Algebra – Selected Topics. IV, 234 pages. 1976.

Vol. 522: C. O. Bloom and N. D. Kazarinoff, Short Wave Radiation Problems in Inhomogeneous Media: Asymptotic Solutions. V. 104 pages. 1976.

Vol. 523: S. A. Albeverio and R. J. Høegh-Krohn, Mathematical Theory of Feynman Path Integrals. IV, 139 pages. 1976.

Vol. 524: Séminaire Pierre Lelong (Analyse) Année 1974/75. Edité par P. Lelong. V, 222 pages. 1976.

Vol. 525: Structural Stability, the Theory of Catastrophes, and Applications in the Sciences. Proceedings 1975. Edited by P. Hilton. VI, 408 pages. 1976.

Vol. 526: Probability in Banach Spaces. Proceedings 1975. Edited by A. Beck. VI, 290 pages. 1976.

Vol. 527: M. Denker, Ch. Grillenberger, and K. Sigmund, Ergodic Theory on Compact Spaces. IV, 360 pages. 1976.

Vol. 528: J. E. Humphreys, Ordinary and Modular Representations of Chevalley Groups. III, 127 pages. 1976.

Vol. 529: J. Grandell, Doubly Stochastic Poisson Processes. X, 234 pages. 1976.

Vol. 530: S. S. Gelbart, Weil's Representation and the Spectrum of the Metaplectic Group. VII, 140 pages. 1976.

Vol. 531: Y.-C. Wong, The Topology of Uniform Convergence on Order-Bounded Sets. VI, 163 pages. 1976.

Vol. 532: Théorie Ergodique. Proceedings 1973/1974. Edité par J.-P. Conze and M. S. Keane. VIII, 227 pages. 1976.

Vol. 533: F. R. Cohen, T. J. Lada, and J. P. May, The Homology of Iterated Loop Spaces. IX, 490 pages. 1976.

Vol. 534: C. Preston, Random Fields. V, 200 pages. 1976.

Vol. 535: Singularités d'Applications Differentiables. Plans-sur-Bex. 1975. Edité par O. Burlet et F. Ronga. V, 253 pages. 1976.

Vol. 536: W. M. Schmidt, Equations over Finite Fields. An Elementary Approach. IX, 267 pages. 1976.

Vol. 537: Set Theory and Hierarchy Theory. Bierutowice, Poland 1975. A Memorial Tribute to Andrzej Mostowski. Edited by W. Marek, M. Srebrny and A. Zarach. XIII, 345 pages. 1976.

Vol. 538: G. Fischer, Complex Analytic Geometry. VII, 201 pages. 1976.

Vol. 539: A. Badrikian, J. F. C. Kingman et J. Kuelbs, Ecole d'Eté de Probabilités de Saint Flour V-1975. Edité par P.-L. Hennequin. IX, 314 pages. 1976.

Vol. 540: Categorical Topology, Proceedings 1975. Edited by E. Binz and H. Herrlich. XV, 719 pages. 1976.

Vol. 541: Measure Theory, Oberwolfach 1975. Proceedings. Edited by A. Bellow and D. Kölzow. XIV, 430 pages. 1976.

Vol. 542: D. A. Edwards and H. M. Hastings, Čech and Steenrod Homotopy Theories with Applications to Geometric Topology. VII, 296 pages. 1976.

Vol. 543: Nonlinear Operators and the Calculus of Variations, Bruxelles 1975. Edited by J. P. Gossez, E. J. Lami Dozo, J. Mawhin, and L. Waelbroeck, VII, 237 pages. 1976.

Vol. 544: Robert P. Langlands, On the Functional Equations Satisfied by Eisenstein Series. VII, 337 pages. 1976.

Vol. 545: Noncommutative Ring Theory. Kent State 1975. Edited by J. H. Cozzens and F. L. Sandomierski. V, 212 pages. 1976.

Vol. 546: K. Mahler, Lectures on Transcendental Numbers. Edited and Completed by B. Diviš and W. J. Le Veque. XXI, 254 pages. 1976.

Vol. 547: A. Mukherjea and N. A. Tserpes, Measures on Topological Semigroups: Convolution Products and Random Walks. V, 197 pages. 1976.

Vol. 548: D. A. Hejhal, The Selberg Trace Formula for PSL (2,ℝ). Volume I. VI, 516 pages. 1976.

Vol. 549: Brauer Groups, Evanston 1975. Proceedings. Edited by D. Zelinsky. V, 187 pages. 1976.

Vol. 550: Proceedings of the Third Japan – USSR Symposium on Probability Theory. Edited by G. Maruyama and J. V. Prokhorov. VI, 722 pages. 1976.

Vol. 551: Algebraic K-Theory, Evanston 1976. Proceedings. Edited by M. R. Stein. XI, 409 pages. 1976.

Vol. 552: C. G. Gibson, K. Wirthmüller, A. A. du Plessis and E. J. N. Looijenga. Topological Stability of Smooth Mappings. V, 155 pages. 1976.

Vol. 553: M. Petrich, Categories of Algebraic Systems. Vector and Projective Spaces, Semigroups, Rings and Lattices. VIII, 217 pages. 1976.

Vol. 554: J. D. H. Smith, Mal'cev Varieties. VIII, 158 pages. 1976.

Vol. 555: M. Ishida, The Genus Fields of Algebraic Number Fields. VII, 116 pages. 1976.

Vol. 556: Approximation Theory. Bonn 1976. Proceedings. Edited by R. Schaback and K. Scherer. VII, 466 pages. 1976.

Vol. 557: W. Iberkleid and T. Petrie, Smooth S^1 Manifolds. III, 163 pages. 1976.

Vol. 558: B. Weisfeiler, On Construction and Identification of Graphs. XIV, 237 pages. 1976.

Vol. 559: J.-P. Caubet, Le Mouvement Brownien Relativiste. IX, 212 pages. 1976.

Vol. 560: Combinatorial Mathematics, IV, Proceedings 1975. Edited by L. R. A. Casse and W. D. Wallis. VII, 249 pages. 1976.

Vol. 561: Function Theoretic Methods for Partial Differential Equations. Darmstadt 1976. Proceedings. Edited by V. E. Meister, N. Weck and W. L. Wendland. XVIII, 520 pages. 1976.

Vol. 562: R. W. Goodman, Nilpotent Lie Groups: Structure and Applications to Analysis. X, 210 pages. 1976.

Vol. 563: Séminaire de Théorie du Potentiel. Paris, No. 2. Proceedings 1975–1976. Edited by F. Hirsch and G. Mokobodzki. VI, 292 pages. 1976.

Vol. 564: Ordinary and Partial Differential Equations, Dundee 1976. Proceedings. Edited by W. N. Everitt and B. D. Sleeman. XVIII, 551 pages. 1976.

Vol. 565: Turbulence and Navier Stokes Equations. Proceedings 1975. Edited by R. Temam. IX, 194 pages. 1976.

Vol. 566: Empirical Distributions and Processes. Oberwolfach 1976. Proceedings. Edited by P. Gaenssler and P. Révész. VII, 146 pages. 1976.

Vol. 567: Séminaire Bourbaki vol. 1975/76. Exposés 471–488. IV, 303 pages. 1977.

Vol. 568: R. E. Gaines and J. L. Mawhin, Coincidence Degree, and Nonlinear Differential Equations. V, 262 pages. 1977.

Vol. 569: Cohomologie Etale SGA 4½. Séminaire de Géométrie Algébrique du Bois-Marie. Edité par P. Deligne. V, 312 pages. 1977.

Vol. 570: Differential Geometrical Methods in Mathematical Physics, Bonn 1975. Proceedings. Edited by K. Bleuler and A. Reetz. VIII, 576 pages. 1977.

Vol. 571: Constructive Theory of Functions of Several Variables, Oberwolfach 1976. Proceedings. Edited by W. Schempp and K. Zeller. VI, 290 pages. 1977

Vol. 572: Sparse Matrix Techniques, Copenhagen 1976. Edited by V. A. Barker. V, 184 pages. 1977.

Vol. 573: Group Theory, Canberra 1975. Proceedings. Edited by R. A. Bryce, J. Cossey and M. F. Newman. VII, 146 pages. 1977.

Vol. 574: J. Moldestad, Computations in Higher Types. IV, 203 pages. 1977.

Vol. 575: K-Theory and Operator Algebras, Athens, Georgia 1975. Edited by B. B. Morrel and I. M. Singer. VI, 191 pages. 1977.

Vol. 576: V. S. Varadarajan, Harmonic Analysis on Real Reductive Groups. VI, 521 pages. 1977.

Vol. 577: J. P. May, E_∞ Ring Spaces and E_∞ Ring Spectra. IV, 268 pages. 1977.

Vol. 578: Séminaire Pierre Lelong (Analyse) Année 1975/76. Edité par P. Lelong. VI, 327 pages. 1977.

Vol. 579: Combinatoire et Représentation du Groupe Symétrique, Strasbourg 1976. Proceedings 1976. Edité par D. Foata. IV, 339 pages. 1977.